AGRICULTURE IN INDIA
Policy and Performance

AGRICULTURE IN INDIA

Policy and Performance

Edited by

B. Sambasiva Rao
Associate Professor
Department of Economics
Nagarjuna University
Guntur (Andhra Pradesh)

SERIALS PUBLICATIONS
New Delhi

First Published–2003

ISBN: 81-86771-09-3

Published by

Head Office : **SERIALS PUBLICATIONS**
4, B.S.S. Hall, Khyber Pass Market,
Civil Lines, Delhi–110054 (India)
Phone: 0091-11-3811659
Fax: 91-11-3812678
E-mail: serials@satyam.net.in

Branch Office : 4830/24, Prahlad Street, Ansari Road,
Darya Ganj, New Delhi-110002 (India)
Phone: 3245225

Typeset By : **SHANKAR BOSE**

Price Rs. : **700.00**

Printed at : **Tarun Offset Printers**

CONTENTS

FOREWORD

Since the very beginning of the planning process in India, high priority has been accorded to development of Agriculture as it supports more than 70 per cent of the people and contributes substantially to Gross Domestic Product of the Country. Agricultural production, which increased substantially after mid-sixties, could not keep its pace in the late nineties. It is due to the decline in the capital formation in agriculture, low investment and lack of infrastructure. Instability and declining factor productivity are the major hurdles of Indian agriculture. The input prices have increased enormously and much faster than the output prices and lack of value addition, agriculture has become less profitable. It is also often mentioned that agriculture has been discriminated against Industry, particularly in terms of public outlays and policies. Agriculture is also confronted with unfavourable terms of trade which affected the allocation of adequate resources and distribution of gains.

Another noteworthy feature is the shift from subsistence-oriented agriculture to export-oriented agriculture and it is possible only if there is adequate investment in agriculture. Many empirical studies indicate that due to constant decline in both public and private investment in agriculture capital formation has been very less. There are clear indications of severe environmental degradation of natural resources, particularly land component in many parts of the country. The increasing pressure of population on land for meeting the requirements of food, fodder, fuel, growing needs of industry in urban areas, improvement in agricultural productivity is necessary.

The New Economic Policy initiated since 1991 has raised certain issues about its implications particularly in agriculture and rural development. The advent of World Trade Organisation (WTO) has added another dimension to overall economy of the country in general and agriculture, in particular. The globalization of the economy is expected to increase more opportunities for export of agricultural products. This might improve the economic conditions of the farmers as they will be able to get better prices for their produce in the global market. In India, more than 80 per cent of the farmers cultivate marginal and small holdings of less than two hectares and find it difficult to meet the cost of farm inputs, particularly purchased inputs. There is a widespread feeling among the farming community that the benefit of globalization will go to large farmers and this will further accentuate the existing disparity in the rural sector. In these circumstances, many issues which need through discussion to realize the full benefit of WTO and take necessary precautionary measures to avoid any problems in the future.

Keeping in view the development of agriculture as the lifeline of the Andhra Pradesh's Economy, the state has massive strengths in agriculture including climate, water resources, soil and a large coast line that facilitates exports. The vision 2020 document of Andhra Pradesh has also highlighted the potentialities of the growth of agricultural sector and stated the state will be a powerhouse of Indian agriculture, with a well developed, efficient and productive allied sector and strong agro-based Industries. Further, to realize the goals of equitable development and growth, the focus needs to develop the backward, drought-prone and rain-fed areas.

With these aspects in mind, the Department of Economics organized a two-day National Seminar on "Agriculture in

Andhra Pradesh: Policy, Performance and Prospects" and as a follow up, attempts to bring out the proceedings. The papers in this volume contain valuable information on Agricultural development and the emerging problems based on the latest field level data in Andhra Pradesh and other states. I hope this volume will contribute literature to the policy makers to a greater extent.

Nagarjuna University
Nagarjuna Nagar
Guntur-Andhra Pradesh

Prof. L. Venu Gopal Reddy
Vice-Chancellor

PREFACE

Even since Independence, India has pursued a deliberate policy to modernize and revitalize its agriculture with the objective of improving the living conditions of the large proportion of the population dependent on agriculture. The policy has consisted of a combination of institutional reforms and large investment in irrigation, power, roads and other rural infrastructure with a view to accelerating agricultural growth. The policy frame work resulted in accelerating the growth rate of agricultural output to about 2.7 per cent per annum (from 1949-50 to 1996-97) as compared with a growth rate of less than 0.8 per cent per annum during the first half of the century.

The introduction of New Seed-Fertilizer technology marks a significant development since it initiated a process of transformation of traditional household agriculture into modern and scientific agriculture in different parts of India. The technological transformation of agriculture was facilitated by large-scale investment in research and extension. The new technology was instrumental in bringing about unprecedented increases in yield levels and output of major crops like wheat and rice. The advent of new technology was also responsible in bringing India out of the huge food shortages of the mid-1960's by raising the growth rate of agricultural and food grains output in India.

The Economic Reforms initiated in June 1991 aimed at putting the economy on a high growth path through improvements in efficiency in the Industrial sector by bringing in competition. Unlike Industries, agriculture has performed more or less in a competitive environment all along. Nevertheless, inefficiencies in resource use in agriculture arose due to trade restrictions, inappropriate pricing of inputs and outputs and the prevailing macro environment. The focus of the New Economic Policy has been to integrate the Indian Economy with the global market. A series of radical economic changes have been initiated aimed at bringing about structural changes (structural reforms) to accelerate growth in different sectors of the economy, particularly the agricultural sector, merits deeper enquiry.

Agriculture in India is still the major sector contributing about 30 per cent of the GDP and providing employment to about 70 per cent of the total working force. The new World Trade Treaty finalized after several rounds of the Uruguay round of negotiation made under General Agreement on Trade and Tariff (GATT) has important implications for the agricultural sector in particular and the national economy in general. The long term objective of the World Trade Organisation (WTO) is to establish a fair and market-oriented agricultural trading system through substantial progressive reductions in agricultural support and protection over an agreed period of time with specific binding commitments.

Keeping in view the policies and performances in agricultural sector and the implications of WTO provisions, the department of Economics, Nagarjuna University conducted a National Seminar on "Agriculture in A.P. Policy, Performance and Prospects" during March 10-11, 2001. This book is a collection of the papers submitted by social scientists and policy makers in the seminar.

For this, I wish to place on record my sincere thanks to Professor L. Venugopal Reddy, Vice-Chancellor, Nagarjuna University, who has taken interest and accepted to write

"FOREWORD" to this Volume. I am also grateful to Professor C.V. Raghavulu, Former Vice-Chancellor, Nagarjuna University for his encouragement and guidance in all my academic activities.

I express my sincere thanks to all my colleagues in the Department of Economics who put in great deal of hard labour and helped in making the seminar a real success. My grateful thanks are due to the paper writers, rapporteurs for recording the discussion. Lastly, I am grateful to Mr. S.K. Jha Serials Publications, New Delhi and his team for processing the manuscript for publication.

Dr. B. Sambasiva Rao

LIST OF CONTRIBUTORS

I.V. Subba Rao, *Vice-Chancellor, Acharya N.G. Ranga Agricultural University, Rajendranagar, Hyderabad.*

S. Iyyampillai, *Associate Professor, Department of Economics, Bharathidasam University, Tiruchirapalli, Tamil Nadu.*

S.K. Peeran, *Reader in Economics, Government Degree College, Vidavalur, Nellore District, A.P.*

B. Raghavulu Naidu, *Professor of Economics, S.V. University, Tirupati.*

D. Harinarayana, *Director, School of Distance Education, Andhra University, Visakhapatnam.*

L.D. Sudhakara Babu, *Associate Professor in Economics, School of Distance Education, Andhra University, Visakhapatnam.*

K Ravi Babu, *Associate Professor in Economics, Andhra University, Visakhapatnam.*

R.G. Desai, *Professor and Chairman, Department of Economics, Bangalore University, Bangalore.*

Abdul Noor Basha, *Professor, Department of Commerce and Business Administration, Nagarjuna University, Guntur.*

V. Badari Narayana Raju, *UGC Research Fellow, Department of Commerce and Business Administration, Nagarjuna University, Guntur.*

B. Chakravarty, *Dy. Director, Centre of Micro Planning National Institute of Rural Development, Rajendranagar, Hyderabad.*

K.B. Reddy, *Professor & Head, Department of Environmental Sciences, Nagarjuna University, Guntur.*

C. Ratnam, *Honorary Director, Agro-Economic Research Centre, Andhra University, Visakhapatnam.*

B. Chinna Rao, *Field Officer, Agro-Economic Research Centre, Andhra University, Visakhapatnam.*

N. Sudhakara Rao, *Professor, Department of Economics, Nagarjuna University, Guntur.*

B. Sambasiva Rao, *Associate Professor, Department of Economics. Nagarjuna University, Guntur.*

Y. Indira Kumari, *FIP Teacher Fellow, Department of Economics, Nagarjuna University, Guntur.*

M. Babu Rao, *Research Scholar, Department of Economics, Nagarjuna University, Guntur.*

N. Vijaya Ratnam, *Professor, Department of Commerce and Business Administration, Nagarjuna University, Guntur.*

A.K.V.S.S.R. Raja Sekhar, *Reader in Economics Hindu College, Guntur.*

M. Raja Sekhara Babu, *Reader in Economics, V.T.J.M. & I.V.T.R. Degree College, Mangalagiri.*

L.K. Mohana Rao, *Professor, Department of Applied Economics, Andhra University, Visakhapatnam.*

P. Narasimha Rao, *Associate Professor, Department of Economics, Nagarjuna University, Guntur.*

I. Satya Sundaram, *Lecturer, Department of Economics (PG), The Hindu College, Masulipatnam.*

M.V. Narasimha Sarma, *Professor, Department of Economics Nagarjuna University Guntur.*

G.B. Franklin, *Teacher Fellow, Department Economics, Nagarjuna University Guntur.*

T. Hanumantha Rao, *Project Fellow, UGC/MRP, Department of Economics, Nagarjuna University Guntur.*

C.S.N. Raju, *Professor Department of Economics, Nagarjuna University, Guntur.*

K. Raja Babu, *Professor, Department of Economics, Nagarjuna University, Guntur.*

M. Koteswara Rao, *Associate Professor and Head, Department of Economics, Nagarjuna University Guntur.*

P. Venu Gopal, *Lecturer, Department of Economics, Hindu College, Guntur.*

N. Satyanarayana, *Senior Lecturer, Department of Economics, P.B. Siddhartha Autonomous College, Vijayawada.*

J. Vasantha Kumari, *HOD, Economics, SDM Siddhartha Mahila Kalasala, Vijayawada.*

R.V. Rao, *Teaching Assistant Department of Economics Nagarjuna University Guntur.*

A. Ranga Reddy, *Professor, Department of Economics, Sri Venkateswarea University, Tirupati.*

L.S.N. Prasad, *Reader in Economics, Hindu College, Guntur.*

P. Vasudeva Rao, *Department of Economics, Hindu College, Guntur.*

R. Siva Ramprasad, *Teaching Assistant, Department of Computer Science & Engineering, Nagarjuna University Guntur.*

A. Sreeram, *Lecturer in Economics A.L. College, Vijayawada.*

1

Agriculture in Andhra Pradesh—A Scientist's Perception

* *I.V. Subba Rao*

From ship to mouth, begging bowl to bread basket and then to marketable and exportable surplus in agricultural production, India has made tremendous progress in agriculture front. In 1798, Thomas Malthus predicted an adverse imbalance between human population and food production and catastrophe. Again in mid sixties, Paddock brothers have virtually written off India and forecasted famine and large scale starvation deaths, India has not only disproved both the theories but became a giant in agricultural production and a strong Nation, despite nearly three fold increase in population from 361 millions in 1951 to 1000 millions by May 2000.

Andhra Pradesh is the fourth largest state in terms of area and population in the country. The State is agrarian in character and agriculture is the life line and main stay of the people. The State is strategically located and is endowed with a wide range of agroclimatic conditions and natural resources conducive for the development of agriculture and allied sectors. Andhra pradesh, in a sense, is a microcosm of the nation having an advanced, prosperous area in the coastal districts, an economically and socially backward area in Telangana, a drought-prone area in Rayalaseema and tribal belt along the northern borders. The average accomplishments of the State cloud these wide disparities. Further, the progress achieved over the periods is often dampened by the recurrent natural calamities like floods and cyclones in coastal districts and prolonged droughts in the rest.

How Important is Agriculture in Andhra Pradesh

- Agricultural Sector accounts for nearly one-third of GDP.
- Provides sustenance to 65 per cent of the population.
- Accounts for about 30 to 40 per cent of exports.
- Contributes 4 millions tonnes of food grains every year towards national food grains buffer stocks.

THE SWOT ANALYSIS

Strengths

- Endowed with a reservoir of natural resources yet to be exploited, varied agroclimatic conditions, abundant sunshine, different types of soils enabling the cultivation of a variety of crops.
- Has farmers of high receptivity, knowledge, business acumen and regaining ability.
- Has comparative advantage in agricultural, horticultural and animal and fish exports.
- Has ideal conditions for large scale commercial production/storage and marketing of quality seeds and plant material for export.
- Has second largest coastline (974 km) in the country.
- Strategically located in south-central part of India, with easy access to all parts within and outside the country.

Weaknesses

- Long coastline-often subjected to cyclones, storms, floods and drainage impedence.
- About 60% of gross cropped area is rainfed.
- Fragmented and small holdings.
- Low level of literacy.
- Precarious power supply.

Opportunities

- Location of several National Institutes, an International Institute and several public and private sector organizations working on agricultural R and D, relevant to the state.
- Has capacity to exploit and encash off-season markets of developed countries for fruits and vegetables.

Threats

- Erratic distribution of rainfall.
- Droughts, floods and cyclones.
- Threat of extremism–ineffective and non or under utilisation of land and other resources.
- Trade barriers (both tariff and non-tariff) for horticulture/ floriculture exports.
- Volatile and unpredictable international prices.

PROFILE

Demography

The population was more than doubled from 361 lakhs in 1961 to 760 lakhs in 2000. The percentage of people depending on agriculture has declined from 68.7 to 65.1 between 1960-61 and 1990-91, although their number increased from 128.2 to 195.2 lakhs. Similarly, the strength of the cultivators decreased from 40.1 to 28.0% during the same period (Annexure 1.) On the other hand, the agriculture labour force more than doubled from 54.4 (28.6%) to 116.3 lakhs (41%). The rural population decreased from 82.6 to 73.1%, while the urban population increased from 17.4 to 26.9%. This was mainly because of migration of people from rural to urban areas for employment and livelihood. Interestingly, people below the poverty line decreased from 48.9% in 1973-74 to 22.2% in 1993-94.

Land Use

Out of the geographical area of 274.4 lakh ha, the net cropped area is only 39.0% (1999-2000) and this has declined from 113.7 to 98.46 lakh ha (1997-98). But, the gross cropped area increased to 136.2 lakh ha (1999-2000) from 124.44 lakh ha (1956-57). This is mainly because of improved irrigation facilities from 28.60 to 43.84 lakh ha over a period of time. Bulk of the irrigation comes from wells (3.21 to 19.0 lakh ha) and canals (12.74 to 16.34 lakh ha). The area irrigated under tanks declined from 11.80 to 6.52 lakh ha between 1955-56 and 1999-2000. The forest area increased from 58.52 to 62.0 lakh ha, but is still below the norms for ecological health. The land put to nonagricultural uses increased from 18.3 to 25.0 lakh ha indicating diversion of valuable productive agricultural land to non-agricultural and non-productive uses. This tendency, if unchecked, will be detrimental to food security for the growing population. The culturable waste land, however, declined from 16.3 to 7.7 lakh ha, a progressive sign of agricultural development. The area under permanent pastures, basic to livestock health and prosperity, has eroded from 12.0 to 6.9 lakh ha. As a result, the animal sector suffered and would continue to suffer if corrective measures are not applied.

The land use statistics indicate a large fragmentation of the land. The number of operational holdings increased from 42.6 lakhs in 1956 to 106.03 lakhs in 1996, the average size of the land holding decreasing from 2.4 to 1.4 ha, a change not conducive for scientific and profitable farming.

Performance of Agriculture Sector

Andhra Pradesh made rapid strides in agriculture sector and made impressive progress in the last 4½ decades. Thus, the state is in a very comfortable position with regard to food security. Besides being self sufficient, it contributes about 4 million tonnes of foodgrains to nation's buffer stock every year.

The foodgrains production more than doubled between 1956-57 and 1999-2000 from 58.9 to 137.0 lakh tonnes, touching a peak of 149.1 lakh tonnes in 1998-99. This is inspite of a decline in the area under foodgrains by about 20% from 89.3 to 71.4 lakh ha between 1956-57 and 1999-2000, because of a 2 to 4 fold increase in the productivity of major food crops (Annexure 2). The productivity gains are 133% in rice, 64% in *jowar,* 483% in maize and 32% in redgram. Four other crops which have shown substantial improvement in production and productivity are cotton, sugarcane, chillies and groundnut.

From the state's economy point of view, the most important crops are rice, groundnut, cotton, chillies, sugarcane, mesta, fruits, vegetables and to some extent maize, turmeric and pulses. Among the recent introductions, the soybean and oilpalm crops are expected to play important role. Even coriander and bengalgram crops have shown promise and their area is on the increase.

Agricultural Growth

Agricultural development between 1950-51 to 1968-69 was area-led, the net sown area increased from 103 to 113 lakh ha (at ACGR of 0.7%). The area irrigated rose from 23.56 to 31.32 lakh ha (1.59%) and double cropped area from 8.0 to 18.0 lakh ha (3.9%).

Between 1968-69 to 1980-81, growth in output was yield-led. Yields of all crops have gone up at an exponential growth rate to varying extends (0.18 per cent in sugarcane to 3.99 per cent in maize).

Barring a few years and in certain crops, yield in the subsequent period did not rise appreciably and agricultural growth reached a plateau. This has been attributed to a decline in public investment in irrigation, the petering out of the impact of the new technologies, whose focus was mainly on irrigated crops. The rainfed crops accounting for over 60 per cent of the cropped area and their production base have not received that much attention.

The annual compound growth rate of agriculture was 2.5 to 3.0 compared to population growth rate of 2.0 to 2.2 per cent. It is, thus, obvious that the state agricultural sector is quite comfortable. This should be improved to 5.7% by 2020.

Trends in area, production and productivity of important crops in Andhra Pradesh and sectoral growth rates are given in Annexure 3.

The agriculture sector growth rate for Andhra Pradesh varied between 2.65 (1990-91 to 1995-96) to 3.85% (1985-86 to 1990-91) which was slightly less than the all India growth rate of 2.35 to 4.11% during the same period. The per capita NSDP growth rate for Andhra Pradesh ranged between 1.46 (1990-91 to 1995-96) and 2.81% (1985-86 to 1990-91) compared to all India average of 2.59 (1980-81 to 85-86) to 3.71% (1985-86 to 1990-91). The agriculture sector in Andhra Pradesh showed an average growth rate of 3.16% between 1980-81 to 1995-96 comapred to all India average of 3.24%

Performance of Andhra Pradesh *vis-a-vis* Other States

A comparison of the food grains production in the major states (Annexure 4) revealed that although the state's foodgrains production on an average rose from 70.41 lakh tonnes for the triennium in 1968-69 to 126.94 lakh tonnes in 1996-99, a few other states in the country surpassed Andhra Pradesh in this respect. Punjab recorded 402% growth compared to 80% of Andhra Pradesh, Haryana, Uttar Pradesh, Rajasthan, West Bengal, Madhya Pradesh, Maharashtra and Bihar registering growth rates of 309, 196, 169, 156, 143, 110, 108 per cent respectively. However it's performance was better than the other Southern states, Tamil Nadu recording 50% during the same period. This is perhaps because of the higher hectarages under foodgrains and /or higher percentages of irrigated areas in these states. While Andhra Pradesh has 40.7% of the cultivated area under irrigation, states like Punjab, Haryana and Uttar Pradesh have 94.1, 78.8 and 66.9% of cultivated area under irrigation.

Position of Andhra Pradesh for Major Crops in India

Annexure 5 gives the status of Andhra Pradesh in the country in respect of area, production and yield of major crops.

Area-wise, Andhra Pradesh ranked one in respect of groundnut, mesta, chillies, turmeric and tobacco, 2 in castor, 3 in sunflower, cotton, coriander, 4 in *jowar,* 5 in *ragi,* redgram, sugarcane and 6 in rice and maize.

Production-wise, Andhra Pradesh stood 1 in mesta, chillies, turmeric, tobacco, 2 in maize, 3 in rice, groundnut, castor, sunflower, cotton, coriander, 5 in *Jowar, ragi,* sugarcane and 6 in redgram.

Yield-wise, Andhra Pradesh ranked 1 in maize, mesta, chillies, 2 in turmeric, 3 in rice, tobacco, 4 in *ragi,* sugarcane, 5 in sunflower and 6 in groundnut, cotton, coriander.

While Andhra Pradesh stands above the national average in respect of the yields of rice, maize, *bajra,* cotton, sugarcane, chillies, mustard and sunflower, it is below the national average in *jowar, ragi,* redgram, groundnut, castor, sesamum and tobacco crops.

The average highest yields reported under various crops in the state are as under:

RECORD CROP YIELDS (HIGHEST) IN ANDHRA PRADESH SINCE 1970-71

Year	*Crop*
1981-82	Sugarcane (83249)
1992-93	*Jowar* (864), Tobacco (1363)
1993-94	Cotton (315)
1994-95	*Ragi* (1320), Castor (342)
1995-96	*Bajra* (969), Groundnut (1183), Sesamum (271)
1998-99	Rice (2812), Maize (3470), Redgram (433), Chillies (2361)
1999-2000	Mesta (1470)

(Figures in parentheses indicate yields in kg/ha)

Compared to the yields reported above in Annexure 2, the yields in the farmer's fields and on-farm demonstrations are much higher. Thus, there is a wide gap between the current yields of several crops and those that can be harvested under on-farm conditions and this has to be bridged and exploited to the advantage of the farmer and the state. Under short-term developmental strategies the state should give priority to this item of work so that with a minimum effort and least investment the crop yields can further be pushed in a short time.

Crop Shifts

In the preceding 4½ decades, *jowar* has lost nearly about 18 lakh ha, while the area under total foodgrains declined by about 18 lakh ha. These areas have been diverted to crops like cotton, pulses and oilseeds, mainly groundnut. The total foodgrains production, however, increased for the reasons already explained. Because of various factors, cotton cultivation extended fast to non-traditional areas and unsuitable soils, inviting associated socio-economic problems. It's area increased from 4 to 12.8 lakh ha (1998-99) and production from 1.3 to 18.8 lakh tonnes (1996-97). The area as well as production of sugarcane have improved. Though the productivity increase is marginal, the sugar recovery has increased substantially. The state continues to enjoy the monopoly in chillies, turmeric and mesta. Some crops, non-traditional to the state, like soybean, and mustard and bengalgram are gaining ground. The increase in the areas of maize and soybean can be attributed to their diversified industrial uses and increasing demand. Crop diversification from food to non-food crops, though right from farmer perspective, needs to be regulated cautiously from food security point of view.

The state occupies a predominant place in the country in the cultivation of fruits, vegetables, spices, oilpalm and flowers. More than 13 lakh hectares are under horticultural crops, with an estimated production of 8 million tonnes. Of this, fruits account for 5.3 million tonnes, vegetables 2 million tonnes and spices about 1.1 million tonnes. It is the third largest producer of coconuts in the country.

Agricultural Inputs

Land, labour, capital, seed, fertilizers, irrigation, pesticides etc. constitute the inputs for agricultural production. The country witnessed Green Revolution mainly through high yielding improved varieties and hybrids of crops, irrigation, fertiliser application and plant protection measures.

Seed

Seed is the most important single contributing least expensive input for realising higher yields. Among the technologies, the rate of adoption of improved varieties/ hybrids of crops is the highest, sometimes approaching 100%. Andhra Pradesh state is foremost in seed production and supply. The public and private sector agencies contributed significantly in this direction. Today, Andhra Pradesh is considered as the Seed State of India, supplying lakhs of quintals of certified seed, both within and outside in several other states in the country. A larger number of seed companies are in operation in Andhra Pradesh, which distribute large quantities of Certified Seed to the farming community. Consequently, the area coverage under high yielding varieties approaches 100% in certain crops. The National Seeds Corporation is also producing large quantities of seeds of crop varieties released by the University and distributing in the other states. The ANGRAU has so far released 268 high yielding crop varieties, which are highly popular in Andhra Pradesh and in 14 other states.

Seed Village concept, launched by the Andhra Pradesh Government and the University is a step in right direction. Under this, the farmers can produce their own seed, at low cost, with the Breeder and Foundation seeds supplied by the University. This will ensure quality, which is quite often a causality in the open market. Production and supply of pure and quality seeds of improved varieties should be given priority in the developmental plans.

Fertilisers

Andhra Pradesh is the second highest fertiliser consumer in India. The consumption which was 2.96 lakh tonnes in 1968-69 has touched 20.07 lakh tonnes in 1998-99. fertiliser use, however, is largely confined to irrigated belt, rainfed areas getting least proportions. Even among the irrigated crops, rice, cotton, sugarcane, chillies, turmeric get the lions share. The per ha consumption was 23.8 kg/ ha in 1968-69 compared to 147.3 kg/ha in 1998-99. However, the most important aspect to be considered is the ratio in which the major nutrients NPK are applied. The NPK ratio which used to be very wide (11.6 : 3.7 : 1 in

1970-71) has narrowed down to 7.9 : 3.4 : 1 in 1998-99, largely due to the use of P and K, because of Governmental subsidies for P in and K fertilisers. There is excessive and indiscriminate use of N fertilizers, and farmers apply P and K to soils, which do not need them Consequently, the cost of production is going up.

Intensive cropping has led to depletions of secondary and micro nutrients, resulting in their deficiencies, depressing the crop yields. Hence, what is required now is the application of fertilisers in balanced proportions on the basis of soil test data and crop needs. To sustain productivity and protect environment and avoid degradation of soil, the Integrated Nutrient Management should be practiced. This includes application of organic manures such as composts, crop residues, green leaf and green manure and bio-fertilisers along side chemical fertilisers. All such practices help to reduce the cost of production as well.

Pesticides

Andhra Pradesh has got the dubious distinction of consuming the largest quantity of pesticides. The consumption of pesticides has gone up to 74.69 lakh litres (1998-99). However, the per ha consumption is 0.85 litres, which is low compared to many other countries. The indiscriminate, excessive, misuse and abuse of pesticides have led to serious problems, besides pushing cost of produce and leaving residues of pesticides on produce in prohibitive levels. These include development of insect resistance to pesticides resurgence of pests, pollution of soil, water and environment, high pesticide residues in agricultural produce, feeds, fodders, animal products including meat, mutton, milk (breast milk being no exception in this regard), fish and fish products, fruits, vegetables, thereby, endangering the health of children, human beings, animals, fish etc. Because of the high pesticide residues in the produce, we are loosing our export markets outside the country. The spurious and adulterated pesticides add to the misery of the farmers. To overcome the above problems, integrated pest management (IPM) and integrated disease management (IDM) have been introduced successfully. Thus, in the recent years the pesticide consumption has gone down by about 57%. The IPM, involving use of resistant crop varieties, botanical and bio-pesticides, pheromone traps and need based chemicals (as a last resort) is the panacea for all these problems. Biological control employing parasites, predators, viruses, bacteria, fungi, pheromones, juvenoids and chitin synthesis inhibitors should be promoted.

OTHER GROWTH INDICATORS

Power

Today, agriculture sector is the single largest consumer of power in the state. The electricity consumption in the agriculture sector was mere 57 million KwH IN 1960-61 accounting for 10 per cent of the total consumption (571 million KwH). This rose to 9799 million KwH in 1997-98 accounting for 40.6 per cent of the total consumption of 24154 million KwH.

Farm Machinery and Implements

Traditionally, human labour is employed for almost all agricultural operations, women accounting for large part of this work force. Apart from non-availability of labour for timely operations, drudgery and, hence, the low efficiency of labour, all add to the increased costs of cultivation. To reduce the cost of cultivation, production and supply of improved implements, and small machinery at affordable costs need to be stepped up. The use of improved implements and farm machinery is on the increase.

Particulars	*1983*	*1993*	*%Increase*
Iron ploughs	246,817	384,118	64
Electric water pumpsets	437,583	933,988	113
Tractors	21450	56887	165
Sugarcane crushers	27343	25685	6

However, complete mechanization of farm operations, replacing rural labour work force, is neither feasible nor should be attempted because of the small size of the individual land holdings and also, we should not replace the rural labour work force with machines. Selective or partial mechanization should be attempted to reduce the labour cost, cope up with labour scarcity, minimize drudgery for women, improve work efficiency, timely conduct of tillage, sowing, planting and other farm operations. Less expensive, hand operated and bullock drawn implements and small farm machines should be made available.

State Domestic Product

The NSDP is Rs. 15586 crores in 1996-97 based on constant prices (1980-81). The contribution of agriculture to NSDP is Rs. 5427 crores accounting for 34% (1996-97) compared to 54% in 1960-61. Based on 1993-94 prices, the NSDP and GSDP for the state are as follows (1998-99):

	GSDP	*NSDP*
Total for the State (Rs. in crores)	75530	74263
Agriculture (Rs. in crores)	19776	19021
Share of Agriculture (%)	26.2	25.6
Per Capita Income (in Rupees)	9118	

The share of agriculture has come down to about 26% by 1998-99, a sign of industrialization.

Capital Formation

The data on gross domestic capital formation in Andhra Pradesh in relation to state income in agriculture are given in Annexure 6. The gross domestic capital formation in agriculture has shown increase from Rs. 194.19 crores in 1980-81 to Rs. 1082 crores in 1993-94. Expressed in terms of percentage of gross domestic capital formation in agriculture to the state income from agriculture, it increased between 1980-81 (5.9) and 86-87 (11.1), but started declining thereafter to about almost half of that, in the year 1995-96 (5.1). The data also reveal that the share of private sector has declined over the same period (67.4% in 1980-81 to 48.51% in 1993-94) where as the share of public sector increases from 36.0% in 1981-82 to 51.5% in 1993-94. This is quite interesting.

The pace of economic development of Andhra Pradesh is significantly influenced by the pace of its agricultural development. The state income from agriculture, including allied activities, at constant prices increased by about 2.3 times, while the state income from all sectors increased by 3.8 times, Aggregate growth performance of the state agriculture, at constant prices, as revealed by the crop and livestock and fisheries sectors combined, was higher in the decades 1980s and 1990s (2.06 and 3.59) over its achievements in the early par (-0.04 and 1.59 in 1960s and 1970s). But, when compared to the national performance, the growth was not that encouraging. The GDP from agriculture at constant prices, including livestock and fisheries sectors, recorded an annual average growth rate of 2.77 per cent in the state during the period 1980-81 to 1996-97, compared to 3.33 per cent at national level. The per capita income from agriculture remained stagnant in the state at around Rs. 2350/- from 1970-71 to 1990-91, where as the same increased from Rs. 3077/- to Rs. 3352/- at the national level.

Labour Wages

Wages to agriculture labour showed constant increase for men, women and non-adults in the last two decades (Annexure.7). The average daily agricultural wages increased by 316, 301 and 290 per cent for men, women and non-adult labour, respectively. On the other hand for a labour intensive crop like rice, the corresponding increase was only 171%.

Exports

Andhra Pradesh has got good potential for export of agriculture, horticulture, animal, fish etc. products. This, however, has not been exploited to full advantage. The high cost of production and low or poor quality of produce, the high pesticide residues in the produce, high afflotoxin contents in certain products are deterrents in this regard. Lack of processing, lack of infrastructure, lack of better transport facilities, lack of cold storage, lack of international airport, low product subsidies, low export subsidies are the other constraints for large scale exports. The exports from Andhra Pradesh state since 1971 are given in the table below:

Exports from Andhra Pradesh since 1994-95

(in lakhs)

Item Description	*1994-95*	*1995-96*	*1996-97*	*1997-98*	*1998-99*
Agricultural and agro-based products	956.02 (27.5)	1810.3 (35.0)	2646.05 (39.04)	2681.6 (38.3)	2735.24 (37.3)
Leather, Animal and Marine products	857.08	911.18	962.41	970.00	950.40
Total	**3477.22**	**5172.30**	**6777.13**	**7002.65**	**7341.00**

Figures in parentheses indicate percentages

Source : Commissionerate of Commerce and Exports

The export of agriculture and agro-based products accounted for 37.3 per cent of the total export value of Rs. 7341.0 lakhs in 1998-99. But this was 27.5% of the total exports worth Rs. 3477.22 lakhs in 1994-95.

Prospects

The analysis made in the preceding sections highlights the potential for future exploitation.

There is large un-exploited yield gap for many crops between what is achievable and what has been achieved. This has to be bridged on priority. Prevention of losses due to pests and diseases during production and storage adds further to the kitty. Other areas which offer promise for further improving agricultural scenario are outlined below.

Land Resource Management

Soil erosion, salinity, alkalinity and drainage impedence constitute the major causes of land degradation in the state. The extent of areas affected by different degradative processes is estimated to be around 37 lakh ha, which need to be brought under use by suitable means. Technologies are available. But concerted action among all stake holders is required.

The state has an extended coastline of 974 km. The coastal sandy soils are characterised by high permeability, low fertility and poor water retention. However, the enterprising farmers of the state are cultivating these soils by harvesting the stored sub-surface water through shallow dugout pits popularly known as *'Doruvus'*. The *Doruvus* are being exploited for drinking water as well. Adoption of "Improved *Doruvu* Technology" developed by ANGRAU scientists can make these soils more productive.

Waste Land Development

Andhra Pradesh is one of the states having a large area of 24.6 lakh ha of waste lands. Specific programmes are needed to harness this resource particularly through horticulture, agroforestry and silvipastural development. Integrated Watershed development programmes are key to bring the waste lands to productive use. People's participation and their active collaboration is, however, essential to make the programme successful.

Irrigation and Ground Water Exploitation

The gross irrigated area of the state stands at 53.04 lakh ha, which represents only 50% of the potential. Future lies in creating more potential and improving the irrigation efficiency of the large, medium irrigation projects and exploitation of minor irrigation sources. Establishment of Water Users Associations, an initiative of the State Government, is one of the right steps in this direction.

The micro-irrigation techniques such as drip, sprinkler and sub-surface pipe irrigation have to be popularised to achieve economy in water use and enhance efficiency of irrigation water. Research on effective and economic use of plastics in irrigation water use need strengthening.

There is urgent need to desilt the minor irrigation tanks and undertake repairs to the breached tanks and distributories.

Ground water potential in the state is estimated at 4.10 million ha metres, of which about 1.0 million ha metres have been exploited. In the regions characterised by low rainfall, the ground water resource has to be harnessed carefully and this should never exceed the recharge capacity. If this is ignored, we may only succeed in converting the drylands into deserts.

Dryland Agriculture

More than 60% of the cultivated area in the State is rainfed. Watershed management is the key approach to improve the productivity of rainfed crops. This should be implemented in an integrated and scientific manner. The *"Neeru-Meeru"* programme can be dovetailed to this.

Crop diversification, rainfed horticulture and farming systems involving horticulture, livestock and fisheries production, should receive priority in rainfed areas. Attention should be paid towards production (both R & D) of fodders in these areas and encourage mixed farming.

Labour and time saving implements and machinery designed and developed by various research organisations need to be manufactured on a large scale by encouraging Agro-Industries and private entrepreneurs and supplied to the farmers.

The future of animal production activities, including dairy and poultry development in the state, depends largely on production of dual purpose crops such as maize, sorghum, soybean and proper utilization of crop residues, agro-industrial by-products such as, fish meal, wastes from aquaculture industry and oil cakes.

Irrigated Crops

Under irrigated conditions, where no other alternative crop to rice can be grown, rice will continue during *kharif*. During *rabi*, where irrigation water is limited, rice can be replaced by ID crops like groundnut, sunflower, soybean, gingelly, pulses and vegetables to some extent, under tank, well and borewell irrigated situations.

IPM and INM

Indiscriminate use of fertiliser and pesticides could not be prevented or avoided despite the well known facts about their ill effects. There is need to minimize pesticide usage by adopting integrated pest management (IPM) and minimize fertiliser usage by going in for integrated nutrient management (INM). Incidentally, both will reduce the cost of production of crops and drive agriculture towards sustainability. Research and development efforts on the production and popularization of non-pesticidal components of IPM such as botanical pesticides, parasites, predators, bio-agents like viruses, bacteria, fungi etc. have to receive greater attention.

INM and IPM should receive high priority.

Diversification of Agriculture

Diversification of agriculture is the key for dynamic, remunerative agriculture and employment generation.

Horticulture related activities, dairying, poultry, sheep husbandry, piggery, apiculture, sericulture, mushroom culture, fish and prawn culture, fodder production etc. are some areas which should receive greater attention.

Mechanised Agriculture

Complete mechanisation of agricultural operations is neither feasible nor desirable, because of the inability of resource poor farmers to afford the costs of implements and machinery. Further, farm labourers also need employment. However, low cost, simple animal drawn and hand operated improved agricultural implements, which have been designed to reduce drudgery and increase efficiency of farm labour, have to be developed and popularised.

To meet energy needs, for carrying out farm operations, particularly of lifting of water for irrigation, solar energy may be harnessed increasingly in future. Similarly, biomass energy can be employed for processes involving drying of farm produce.

Exploitation of Bio-Technology

The future lies in biotechnology, be it in the breeding and development of stress resistant field or horticultural crops or livestock production, therapeutics, vaccine development, prevention of crop losses etc. But, this is a cost intensive technology which requires adequate financial support from Government and through private participation. The Scientists concerned need wider exposure to basic sciences like molecular biology and bio-chemistry for harnessing the potential.

Natural Calamities and Disaster Management

The state, and the coastal districts, in particular, are subjected often to the furies of droughts, floods, gales and cyclones. The economy of the state is crippled by these calamities regularly. A permanent mechanism has to be set up to tackle the situation arising out of such events. Contingency plans are being drawn regularly, though on adhoc basis, which include strategies and technologies to deal with different calamities.

Food and Nutrition Security

The additional requirement for food for the growing population has to come from the currently cultivated areas and shrinking land and water resources and declining factor productivity.

Improving the productivity of cereals, millets, pulses and oilseeds while maintaining their area at the current level, value addition and popularisation of the use of value added products will ensure food and nutrition security.

Fruits, vegetables, mushroom form a major source of better nutrition. Horticulture, particularly rainfed horticulture, offers immense scope in this regard.

Agro-processing and Marketing

One of the primary aims of agricultural development is to enhance farm productivity and ensure remunerative prices to the farmers.

The whole gamut of issues related to marketing of agro-based raw material and finished goods need revamping. A large proportion of farmers would continue to be exploited by unscrupulous middle men and market forces. Modernisation of market yards, creation of storage facilities and access to reliable market information and market intelligence harnessing information technology, go a long way in reducing the marketing problems of farmers.

Agro-processing, by and large, has not developed to the desired extent. The technology for agro-processing has remained as more of an art and not much emphasis has been laid on technology upgradation. This warrants identification of export-import needs and efforts should be made to upgrade the technologies and widen the technological base. For increased profitability through farm related activities, agro-processing, value addition, cold storage facilities and exports are the major thrusts which can no longer be ignored in R & D efforts.

The benefits of quality improvement and value addition through agro-processing have eluded the farmers. The new economic policy and WTO agreements open a vast arena for global marketing of agri-goods and agro-processed products. The opportunity can be exploited if only a strong link is established between the primary producers i.e the farmers, processors and global markets. There is also enormous loss of perishable produce (35 to 100%) due to lack of proper transport, storage, processing and marketing facilities. The loss is estimated to be of the value of several million rupees in the whole country. The country is the second largest producer of fruits and vegetables and Andhra Pradesh stands first or second in the production of horticultural crops. There is, therefore, a vast unexploited potential for these crops and for agro-processing.

Export Oriented Agriculture

Andhra Pradesh has immense opportunities for export of agriculture, horticulture products, milk and milk products, meat, mutton, beef, pork, hides, skins and textiles. The export oriented agriculture and horticulture can thrive well through some form of contract farming with minimum support/interference and production of quality produce. For purposes of export, it is preferable to concentrate on limited number of crops where the State has adequate potential and comparative advantage over others, as for example export of non-basmati quality rice, mango, chillies, cotton, grapes, oil extractives, aquaculture products, poultry products and meat and mutton. Location specific research support for post-harvest handling of these need to be strengthened to generate technologies to suit to the specifications of importing countries.

Women in Agriculture/Gender Issues

There is an increasing realization of the role of women in several farm related activities. Farm women, therefore, have to be targeted specifically to meet their training needs and technology transfer.

POLICY

The Draft Agricultural Policy of the State Government Envisages

The Government to play the role of a facilitator, encourage private investment, strengthen extension to reach large number of farmers, crop diversification, provide market information and market intelligence using IT, establishment of processing industries, prevent further degradation of land, develop wastelands, bring an additional area of 12 lakh ha under irrigation, give thrust to rainfed agriculture further boost, fillip to develop and popularize the time saving agricultural implements, promote Integrated Nutrient Management, Integrated Pest Management, Integrated Watershed Management and Integrated Water Management, encourage organic farming, increase crop intensity, ensure supply of quality inputs, initiate measures for the establishment of strong industrial base, make efforts for implementing pragmatic crop insurance policy, promote Millet/cereal based industrial parks and Horticultural parks, endeavour to give more crop loans to farmers by banks and cooperatives, distribute Kisan Cards to all farmers, establish and strengthen District Research Stations, make efforts to undertake steps to increase the productivity of all crops, create a market stablisation fund of Rs. 100 crores, encourage contract farming and appoint Agricultural officers on contract basis, encourage voluntary land consolidation, make use of KU and C Bands, revive Rytubandu Padakam, strengthen Rytu Bazaars, remove movement restrictions of all agricultural commodities, establish State and District Agricultural Promotion Committees, organize seed village programmes, form Rytu Mitras, and prepare farmers to face WTO implications through awareness etc.

WTO and its Implications

India was one of the 22 founder Members of the General Agreement on Trade and Tariff (GATT) and signed the GATT agreement in April 1994. The agreement came into operation on 1st January 1995. Many people express concerns and apprehensions fearing adverse impact of GATT agreement on Indian agriculture. The present day ills of agriculture and crash in the prices are being attributed to World Trade Organisation (WTO). According to Kofie Annan, Secretary General of the United Nations, such fears are largely because "the implications and consequences of WTO are not fully understood by many". There are advantages and equal number of disadvantages too. But, the WTO agreement provides opportunites for the Member Countries for marketing their agricultural produce and products globally. This facility can not, however, be exploited by developing countries, because the developed and developing countries do not have the same 'level field' for operations. The developed countries have several advantages, which they can exploit under the guise of "Green Box" and "Blue Box" to their advantage.

In the WTO agreement, there are 3 direct provisions related to agriculture viz., reduction in trade barriers, increased market access and a reduction in the aggregate measure of support (AMS). Each Member Country is required to replace all types of non-tariff barriers and go in for tariffication. They are also reequired to reduce the level of tariff in a time frame. While the industrially developed countries are required to reduce their average tariffs by 36% in six years (by January 2001), the developing countries have to do so by 24%, over the base period average (1986-88). There is another stipulation requiring the Member Countries to provide 3 to 5% market access for agricultural produce (3% in case of developing countries and 5% in case of developed countries). As far as India is concerned, this rule does not apply for the present because the quantitative restrictions (QRs) are there in place, for reasons of Balance of Payments. Press reports, however, indicate that the GOI proposes to remove QRs in respect of 825 commodities, for which QRs were there earlier, w.e.f. 1-4-2001. This provision of minimum market access being established by Member Countries will create wider international market for our goods. Exploitation of this potential is possible only if our agricultural goods are competitive in prices and qualitatively good, meeting the specific requirements of the importing countries. Hence, the country and the State have to identify areas, crops, varieties and agricultural commodities which have comparative advantage in the international trade. According to WTO Agreement, a reduction in support of agricultural production, marketing and income is envisaged. The aggregate measure of support (AMS), which is sum total of all support measures provided by the Member Governments, should not exceed 10% of the total value of agricultural products in case of developing countries and 5% in respect of developed countries. Currently, the AMS in India is far below 10% and, hence, there is no need of apprehension in this regard, since GOI can increase the AMS up to 10%. Such of those incentives and subsidies which have been included in "Green Box" and "Blue Box" are excluded from the calculations to arrive at AMS.

Also, in the WTO Agreement, there are two indirect measures, one relating to Trade Related Intellectual Property Rights (TRIPs) and the other to Trade Related Investment Measures (TRIMs). Under TRIPs, the Member countries have to set up international standards for protection for their innovations, particularly patenting of seeds. Difficulties may arise in the use of patented seeds.

But, protection for plant varieties will have to be provided through an effective *sui-generis* system. This is particularly required since Indian farmers save the seeds of improved varieties for resowing in subsequent seasons. Thus, provisions have been made in the proposed Bill on "Protection of Plant Varieties and the Farmers Rights Bill 1999", yet to be passed by the Parliament. We will be wiser if such a *sui-generis* system is legislated immediately before the WTO-TRIPs Council will take up review of this system.

The apprehensions in India are that developed countries may dump their agricultural goods into our country, which will adversely affect our market and result in prices crash. This can be prevented by use of the restrictions stipulated in the agreement under which the affected countries can invoke anti-dumping measures by way of increasing import duties, subject, however, to the bound tariffs prescribed therein.

Epilogue

What is required now is, more income to the farmer, not merely more productivity and production, and more employment for the rural people. Mere dependence on crop production alone does not contribute to this. Agriculture is seasonal which may be 90 to 100 days in rainfed areas extending to about 200 to 250 days in irrigated areas and 300 days in more than a century old Krishna and Godavari deltas. We should think of gainful employment for farmers and agricultural labourers throughout the year. This is possible by diversification of agriculture to horticulture, agro-forestry and by dovetailing agricultural activities with such other avocations like dairying, poultry, sheep farming, fish culture, bee culture, sericulture, mushroom culture etc. and establishing agro-based and processing industries in rural areas where raw materials are available. There is immense potential for exports of agricultural products, fruits, vegetables, flowers, aromatic and medicinal plants, milk, milk products, meat, mutton, beef, pork, hides and skins, animal wastes, fish and prawns, their products, textiles and garments. Andhra Pradesh should be able to capitalize this potential. But, this requires capital investment and creation of infrastructure like roads, warehousing, cold storage, refrigerated transport, un-interrupted power and water supply, and better marketing facilities. Women can gainfully be employed in activities like processing, value addition, by-product utilization, packaging etc. Cottage and rural industries will play key role in this regard. A paradigm shift from sustenance agriculture to commercial agriculture is required.

Also, there is need for a re-look into land reforms and tenancy laws. Some of the existing measures, provisions, rules, laws and Acts which are age old or prevent sale/ marketing of processed agricultural goods have to be amended in favour of the farmers. Restrictions on inter-district and inter-state movement of agricultural produce have to be removed.

Vital Statistics Showing the Performance of Andhra Pradesh

Annexure—1

	1956-57	*1999-2000*
1	**2**	**3**
Total Geographical area (lakh ha)	274.4	274.4
Gross cropped area (lakh ha)	124.44	136.2
Per capita gross cropped area (ha)	0.34	0.19
Net cropped are (lakh ha)	113.74	107.1
Total food grains (lakh tonnes)	58.92	136.97
Per capita availability of food grains (kg/year)	164	206
Cropping Intensity	109	122

(Contd...)

1	2	3
Net irrigated area (lakh ha)		
Canals	12.74	16.34
Tanks	11.80	6.52
Wells	3.21	19.00
Other sources	0.85	1.98
Total	**28.60**	**43.84**
Irrigation Intensity	117 (1960-61)	131
	1960-61	**1990-91**
Agricultural workers (lakhs)	128.2	195.2
Percentage to total workers	68.7	65.1
Cultivators (lakhs)	74.8	78.9
Percentage to total workers	40.1	28.0
Agricultural labour (lakhs)	53.4	116.3
Percentage to total workers	28.6	41.0

Growth in the Essential Commodities

Annexure—2

(Area in lakh ha; Production in lakh tonnes; Yield in kg/ha)

Crop	*1956-57*	*1998-99*	*% Change over 1956-57*	*1999-2000*	*% Change over 1956-57*
1	**2**	**3**	**4**	**5**	**6**
Rice					
Area	29.3	43.2	47	40.1	37
Production	34.1	118.2	248	106.4	212
Productivity	1163	2812	142	2710	133
Jowar					
Area	25.1	7.5	–70	7.4	–71
Production	11.1	5.23	–53	5.4	–52
Productivity	441	695	58	728	64

(Contd...)

1	2	3	4	5	6
Maize					
Area	1.9	4.0	110	4.5	140
Production	1.1	13.8	1154	14.7	1302
Productivity	557	3470	523	3258	483
Redgram					
Area	1.6	3.94	146	4.3	173
Production	0.4	1.7	325	1.6	260
Productivity	272	433	59	358	31.6
Total Pulses					
Area	13.3	15.9	20	16.5	23
Production	2.8	8.27	195	8.0	190
Total Foodgrains					
Area	89.3	73.7	–17	71.4	–20
Production	58.9	149.1	153	137.0	132
Groundnut					
Area	13.4	21.6	49	18.0	34
				22.2	66
				(1995-96)	
Production	12.2	21.6	77	10.9	–11
				26.3	115
				(1995-96)	
Productivity	914	1082	18	607	–33
				1183	23
				(1995-96)	
Castor					
Area	3.4	2.0	–40	2.6	–22
Production	0.7	0.5	–36	0.6	–9
Productivity	190	224	18	221	16

(Contd...)

1	2	3	4	5	6
Oilseeds					
Area	20.9	27.3	13	25.5	22
				31.4	50
				(1995-96)	
Production	13.8	24.5	78	13.6	–1
				30.3	121
				(1995-96)	
Cotton					
Area	4.0	12.8	220	10.5	160
Production	1.3	15.2	1069	15.8	1105
Productivity	57	202	254	257	351
Sugarcane					
Area	0.8	2.1	162	2.3	200
Production	50.6	165.0	226	185.1	206
Productivity	65831	77231	17	79963	21
Chillies					
Area	1.8	2.2	22	2.7	50
Production	1.2	5.3	342	5.0	317
Productivity	643	2361	267	1852	188

Source : Directorate of Economics and Statistics, Govt. of A.P., Hyderabad

Annexure—3

Trends in Area, Production and Productivity of Important Crops in Andhra Pradesh

Crop	*Exponential growth rates (%)-1949-50 to 1995-96*		
	Area	*Production*	*Productivity*
Rice	–1.01	3.4	2.41
Jowar	–1.81	0.08	1.32

(Contd...)

Crop	*Exponential growth rates (%)-1949-50 to 1995-96*		
	Area	*Production*	*Productivity*
Bajra	–2.62	-0.98	1.30
Maize	1.76	5.74	3.99
Ragi	–1.48	–0.75	1.05
Bengalgram	–2.57	–1.82	–0.87
Redgram	–1.51	0.84	–
Total foodgrains	–0.81	2.40	2.80
Groundnut	–1.80	2.27	0.50
Sesamum	–1.55	–1.56	0
Cotton	1.17	4.72	3.54
Chillies	0.88	3.53	2.67
Sugarcane	2.35	2.51	0.18
Tobacco	0.29	1.61	1.37

Sectoral Growth Rates (%) Andhra Pradesh and All India at 1980-81 Prices

S. No.	*Industry*	*1980-81 to 1985-86*	*1985-86 to 1990-91*	*1990-91 to 1995-96*	*1980-81 to 1995-96*
1.	**Agriculture**				
	Andhra Pradesh	2.98	3.85	2.65	3.16
	All India	3.26	4.11	2.35	3.24
2.	**Allied Agricultureal Activities**				
	Andhra Pradesh	–2.95	4.77	5.20	2.27
	All India	2.85	4.89	6.38	3.60
3.	**Total NSDP**				
	Andhra Pradesh	12.86	18.37	14.76	15.31
	All India	5.04	6.28	5.26	5.52
4.	**Per Capita NSDP**				
	Andhra Pradesh	2.34	2.81	1.46	2.20
	All India	2.59	3.71	2.98	3.09

*All India growth rates pertain to gross domestic product

Note: Growth are compound gorwth rates for two points

Source: Directorate of Economics & Statistics, Govt. of A.P., Hyderabad

Annexure—4

Food Grains Production in Major States

States	*Average production (in '000s tones)*		*Growth*	*Net cultivated area*	*Irrigated area as % of net cultivated area*
	1965-68	**1996-99**	**(%)**	**(lakh ha)**	**(%)**
Punjab	4359	21868	402	41.39	94.1
Haryana	2847	11640	309	36.15	78.8
Uttar Pradesh	13988	41373	196	174.75	66.9
Rajasthan	4931	13268	169	167.9	32.6
West Bengal	5522	14159	156	—	—
Madhya Pradesh	7783	18883	143	197.94	25.8
Maharashtra	5866	12340	110	178.76	14.5
Bihar	6636	13806	108	73.37	46.0
Andhra Pradesh	7041	12694	80	106.37	40.6
Tamil Nadu	5600	8391	50	54.86	51.8

Source: V.Viswanathan, Hindu 22/6/2000 (cols. 1 to 4)

Annexure—5

Position of Andhra Pradesh in India for Major Crops

Crop	*Area*		*Production*		*Yield*	
	1988-89	*1998-99*	*1988-89*	*1998-99*	*1988-89*	*1998-99*
1	**2**	**3**	**4**	**5**	**6**	**7**
Rice	6	6	1	3	3	3
Jowar	4	4	4	5	10	7
Maize	7	6	6	2	3	1
Ragi	6	5	6	5	4	4
Redgram	5	5	9	6	13	10

(Contd...)

1	2	3	4	5	6	7
Food grains	10	7	4	4	10	5
Groundnut	1	1	2	3	12	6
Castor	1	2	2	3	8	8
Sunflower	3	3	3	3	3	5
Oil seeds	*	4	*	5	*	8
Sugarcane	5	5	5	5	4	4
Cotton	5	3	7	3	10	6
Mesta	1	1	1	1	3	1
Chilies	1	1	1	1	1	1
Turmeric	*	1	*	1	*	2
Coriander	*	3	*	3	*	6
Tobacco	1	1	1	1	7	3

* Not Available

Source: Directorate of Economics & Statistics, Govt. of A.P., Hyderabad

Annexure—6

Gross Domestic Capital Formation in Agriculture in A.P.

Year	*Gross Domestic capital formation in agriculture (Rs. crs)*	*State Income from agriculture (Rs. crs)*	*% of col. 2 to col. 3*	*Percentage share in GDCP in agriculture of*	
				Private Sector	*Public Sector*
1	2	3	4	5	6
1980-81	194.19	3314.86	5.9	60.2	39.8
1981-82	198.69	4128.94	4.8	67.4	32.6
1982-83	315.02	4145.12	7.6	46.5	54.5
1983-84	288.95	5137.59	5.6	43.9	66.1
1984-85	388.25	4911.23	7.9	43.2	66.6
1985-86	444.03	5018.22	8.8	46.1	53.9

(Contd...)

1	2	3	4	5	6
1986-87	552.24	4991.35	11.1	44.3	65.7
1987-88	574.45	6412.53	9.0	48.6	51.4
1988-89	649.84	8457.77	7.7	50.8	49.2
1989-90	629.2	9409.88	6.7	51.3	48.7
1990-91	691.85	11531.37	6.0	55.1	44.9
1991-92	898.00	14368.21	6.3	56.9	43.1
1992-93	930.75	14179.02	6.6	48.5	51.5
1993-94	1.082.32	17294.51	6.3	48.5	51.5
1994-95	NA	NA	5.9	NA	NA
1995-96	NA	NA	5.1	NA	NA

NA: Not Available
Source: DES, Capital Formation in Andhra Pradesh

Annexure—7

Average Daily Agricultural Wages in Andhra Pradesh

Year	*Wages in Rupees and Paise/8 hours/day*		
	Man	*Woman*	*Non-adult*
1980-81	5.86	4.42	3.56
1982-83	7.94	5.96	4.65
1983-84	11.09	8.3	6.97
1987-88	13.26	9.62	7.97.
1989-90	16.98	11.99	9.47
1991-92	18.35	13.11	10.49
1993-94	24.35	17.73	13.88

Source: Season and Crop Reports, Govt. of A.P., Hyderabad

2

Changes in the Landholding Patterns in Thanjavur Villages during 1980s and 1990s

S. Iyyampillai

This study is based on the evidences available for two villages in the former Thanjavur district at two different points of time, the difference being about 18 years of vast changes in the economic policies adopted by Tamilnadu State and the central government of India. First survey was made in 1981-82 and the second survey in 1999-2000. The information have been collected from all the households physically found earlier and now. The numbers of households are however different then and now due to in-migration, out-migration and partition in the families. There were 265 households in one village and 290 households in the other village during the first survey. This has now increased to 293 and 334 respectively.

This study has been set in the following manner: First, the land ownership pattern, on the basis of land-size distribution, is shown at two points of time, indicating the absence of any significant emergence of polarization effect-neither on the basis of caste nor class. Second, the village level changes during the study period are traced in such a way to indicate as to how they have gradually influenced the agricultural sector in general and rural households investment preferences in particular.

Evidences available in the study villages suggest that the changes occurred in and around the villages have sharpened the market penetration in village economies. As a result, the villagers have become more attuned to the economic forces than the social and/or cultural forces. Due to these changes, more importance is now attached to the expected economic returns from the lands than the social values of the land ownership. The expected economic returns are found to be now explained more by the control over the market rather than the productivity. This has made the usual analysis of structural changes in terms of land-size less applicable. The evidences available in the villages suggest that the land area owned by the rich has shrunk, but it does in no way mean that they have become poor. The same remark holds good for all groups of landowners, of course. This conclusion could be safely extended to some of the neo-landless who are no worse off than their earlier landowning situation is compared.

Palamputhur

This village is a backward caste (BC) and small farmers dominated village. Earlier the lands were mostly cultivated by shallow wells by using human and animal power to draw water. Since the village is located in the tail-end of the Cauvery river, some amount of water was available for irrigation purposes in the up-lands, where the wells also helped irrigation. The low-lying lands were irrigated by system-tanks which were fed by a canal from the river Cauvery. Earlier, paddy was cultivated in the low-lying lands while mixed low yielding crops were cultivated in the up-lands. Hence the tank irrigated lands valued more as compared to the up-lands. But now the situation is different. The up-lands are irrigated by bore-wells (government and private); high yielding, modern commercial crops (like soya bean) are raised now. Also the up-lands are sutiable for construction purpose, as many new houses are built in those lands, since they are closer to road. As a result, the price ratios of the lands are now just opposite to the earlier position. The major changes in landholding patterns in Palamputhur are the following:

1. The degree of landlessness has increased. It is true for both the backward and scheduled castes (SC). This is however lower for backward caste households and higher for scheduled caste households both then and now.
2. The reduction in landholding size is reported for all size-holders and for both caste groups. This is explained at least by three factors. One, divisions in the family, but this is very small. Two, sales of lands to the next lower groups. Three, conversion of cultivated land into residential uses. The big farmers are found to have sold away the far-off tank irrigated and low lying lands, for which the expected productivity is lower. The lands closer to roads and bore-wells are in greater demand, as their estimated productivity is higher.
3. Very few-land owners (5 out of 218) and many new-landless (34 out of 218) are found. The same trend is seemed in the operation status too. Both castes groups are found in the new landowners and new-landless. Majority of the new landless is however found to have come from the earlier marginal and small farmers groups.
4. Excepting the fact that some big farmers have been able to raise the total market value of landed assets (though there is small reduction in the land area), there is no systematic polarization-neither in the class nor in the caste lines is found. This is mainly because of the fact that the agriculture is now

assigned lower importance as compared to the position gained by non-farm investments in the recent years. The income from non-farm activities has also reduced the Gini concentration ratios for household income and expenditure.

Vengidengal

This village is different from Palamputhur in many respects. The lands of this village are irrigated for generations. Earlier the lands were largely owned by the landlords from the upper castes. When the members of these families could get government jobs with better income and political power, they preferred to migrate to urban centres and became absentee landlords. Depending on the situation, they made arrangements to cultivate the lands either through paid agents or through tenants. A few upper caste landlords stayed back in the village and cultivated the lands, but their family members were really not interested in it. About 80 per cent of the village households are landless SC. They worked as servants and agricultural labourers by generations. Some of the BC households were also found working as agricultural laboures, tenants and marginal farmers. Mostly paddy was cultivated. After the paddy harvest, grams were cultivated. Now the situation has slightly changed. Due to water problem, grams were cultivated. Now the situation has slightly changed. Due to water problem the cropping duration has come down along with the fall in productivity per acre. Hence the absentee landlords were looking for somebody who would be able buy their lands. The native SC households were not capable of buying the lands, rather they were dreaming to get lands out of land reform measures. In this situation a voluntary agency called Land for Tillers Freedom (LAFTI) came to this village. They bought about 150 acres of land from the landlords to sell them to the native people at the price of Rs. 12,000 per acre. For SC households they offered the lands at the price of Rs. 7000 per acre. This made some changes in the landholding patterns in this village. The major land-related changes in Vengidengal are the following:

1. The degree of landlessness has come down, while the land ownership has gone up. This has happened in both the caste groups. This process does get reflected in the poverty reduction. The change, however, does not seem to be due to the individual initiative, but largely due to the efforts made by a voluntary agency called "Land For Tillers Freedom" in one of the hamlets (Porkalakudi) of this village.
2. Some households have also totally lost their land ownership (UC-7, SC-14) and operational status (UC-5, SC-13) again found in both the caste groups. However, while those upper caste households have managed to get income from non-farm sources, the SC households are unable to do so.
3. Number of land operating households and the area of operational landholding size have gone up. Again this has happened in both the caste groups.
4. The number of native farmers who have substantially lost their landholdings is small, as the larger area of land that has now come to the hands of native farmers, belonged earlier to the absentee landlords.
5. Most of those who have substantially increased their land ownership status are found in the scheduled caste and small farmer categories.
6. The degree of tenancy has gone up, however, without any significant change in the form of tenancy. The fixed proportion (50:50) of crop share is the most prevalent form of tenancy.
7. On the whole, the households of upper castes have substantially lost (excepting a very few) their land ownership status, while the land ownership status

for scheduled caste households has gone up. The dominance of absentee landlords belonging to upper castes in the village economy has also more steeply fallen down, which however in no way means that their economic condition has become worse. Some portions of the lands of absentee landlords have gone to the native farmers.

8. Increase in Gini concentration ratio is found for land ownership status. This is because of addition of marginal farmers and hence the increase in the size of farmer population at lower brackets but not so much increase in the land area owned by them.

The general decline in the demand for land particularly from the big farmers is observed and is explained by the relative decline in the remunerative capacity of the agricultural sector.

The differences between the study villages in the above-explained land related changes are found to be mainly expalined by the village level, historical socio-economic structures.

As the evidences show in both the villages, the big farmers are not very much interested in ever enhancing their land holding size. Rather, they are interested in enhancing their total worth of assets by properly choosing the investment ventures. Hence, the lands with better market-oriented facilities alone are give importance particularly in Palamputhur that too after analysing the possibility of investing the money in non-farm sector. The poor ones, who never had the privilege of owning lands, now get some opportunity of purchasing lands without really bothering very much about the quality and/or the net earnings of the lands.

Village Level Changes

As to how the macro level policy oriented changes have affected the agricultural sector in general has been very well explained by the earlier studies (C.H. Rao 1994, V.M. Rao 1994, Ashok Gulati and others 1994, 1995 and 2000, Mehta Rajesh 2000, Acharya S.S 1997). The micro level changes that are results of both the macro level changes and the village level structural factors have not been systematically analysed to see as to how they, in combination influence the status of agricultural sector and hence the land holding patterns. It is in this respect the Palamputhur and Vengidengal village economies present some very interesting insights into the actual problems that are confronted by the rural sector.

In Palamputhur, the following areas have undergone significant changes. Generally there is a feeling that net surplus from agricultural operation has sharply declined while the operation itself has become more skill oriented and a tougher task. On the cost side, labour market has become more commercialized now and gone beyond the control of the farmers. The cheap and efficient family labour, exchange labour and labour from the scheduled caste and labour from the dry villages are now scarcely available. Also, (the labour has become now costly due to (1) the breakdown of traditional food for wage system (2) increased and multiple skill of the labourers whose opportunity cost is relatively greater. Reduction in cattle population along with the non-availability of cheap labour has caused a decline in the use of local farmyard manure, in the place of which now the use of purchased chemical fertilizers has gone up. The change in crop pattern has also let to increase in the use of pesticides. The changes in crop pattern have also caused purchase of seeds from the industrial corporate sector. Besides them, the price of water has also gone up. These factors have caused a outflow of large money from village economies. As a result, the farmers are forced to earn more money and hence to sell more or to produce those items that can be marketed easily. The following have lured the farmers to depend more and more on the market for revenue. They are: (a) the free availability of rice in the open market and through Public Distribution System at lower prices, and (b) attractive price offered for the paddy by the Direct Procurement Centers. Development of infrastructure has expedited this process. The proportion of paddy produced and consumed locally has fast shrunk while the proportion of rice purchased from PDS has increased. Since, the farmers are highly unorganized and divided on many lines, their bargaining capacity is lower and hence, their dependence on market (for both the purchase of inputs and the sale of outputs at the prices fixed at both points by the external forces) can hardly them. Thus, the agricultural operation is losing its importance. Hence, it is not a surprise that the land in Palamputhur, the major agricultural input is also being given lower priority in the investors' preferences.

In Vengidengal the situation is different, though the changes observed in Palamputhur also, to some extent, hold good. Here, the large farmers in the village, and also the absentee landlords of upper castes owning more than 80 per cent of the village lands have started showing interest in selling away all their lands, resulting in price crash. They earlier led very prosperous life in the village, controlling, exploiting and enjoying all the village resources land, water and human labour. However, of late due to land reform measures in the 1970s and general rise in the awareness among those depressed sections in the 1980s, the rural life of landed gentry became relatively difficult. But on the other hand, the expansion of bureaucracy and government agencies in the urban centers became very lucrative options for them. Both these push and pull factors worked together resulting in the gradual out migration of rich landlords and their family members. On the other hand the chromic poverty of the landless scheduled caste households was always acting as a hindrance at all points, and hence, though they tempted to own some lands, they could not pay the price bid by landlords. This situation invited the LAFTI to enter into the scene in on e of the hamlets (Porkalakudi) of Vengidengal revenue village. LAFTI could make some people landowners and some others to raise their land holding size. This is the brief history of change in the landholding structure of the village. In this village, neither the market penetration nor the infrastructure development is found to have played any significant role in the agricultural business. On the whole agricultural productivity per unit of land is found to have declined, which can be atrributed mainly to the unfavourable land ownership pattern, through increased lack of water availability is also partly responsible. In the agricultural sector no substantial measure to improve the productivity has been taken—neither by the individuals nor by the government/panchayat. The predominance of paddy cultivation continues. Some efforts to make use of underground water of annual and perennial crops have also proved futile. Under such situation though there are many who are ready to sell the lands (absentee landlords and LAFTI) there are not so many buyers. The LAFTI has a bitter experience of very poor recovery of the money from the new landowners. Even the subsidy (about 40%) provided by LAFTI to the scheduled caste households has failed to attract the buyers.The new landowners do not seem to have significantly improved their economic condition. As the new landowners are unable to mobilize money for meeting out the increased cost of cultivation, they are caught in the hands of moneylenders. Many new farmers find the revenue from lands as inadequate to ever repay their loans. Beside these problems, the village is yet to get potable water. Hence, those who could manage to move outside this village are happy to do so. Only those who have no other go stay back in this village.

Conculusion

The experience of the study villages shows that the land ownership may fail to explain the economic status (leave alone social and political status) of the households. Increase or decrease in landholding size does not necessarily associate itself with the economic status of many households. The correlation coefficient between the landholding size and the household income shows that only 40 per cent of the household income is explained by the landholding size. As long as the agricultural was the mainstay of the households, land was considered to be every thing; the land assumed social, political and economic signigicance. Now it is not so because, land itself cannot be productive (as it used to be), unless it is available in a package with other facilities (proximity to transport, water and markets). Due to these reasons, the rich landlords give lower priority to accumulate land assets (or they are very choosy) and hence, the poor ones and scheduled people are also able to own land; but this does not necessarily mean that their economic and social status in the village go up. It is akin to (economic) sanskritisation process where in the former land owners leave the lands and pick up better options whiel the landless want to become landowners.

REFERENCES

Acharya S.S., 1997, Agricultural Price Policy and Development: Some Facts and Emerging Issues, Presidential Address, 56th annual Conference of IJAE, Thrissur, Kerala.

Gulati Ashok and Sharma, 1994, Agriculture under GATT: What it Holds for India, EPW, July 16, pp. 1857-1863.

Gulati Ashok and S. Bhide, 1995, What do the Reformers have for Agriculture?, EPW, Vol. 31, No., May 6-13, pp. 1089-1093

Gulati Ashok and Narayanan Sudha, 2000, Demystifying Fertilizer and Power Subsidies in India, EPW, Vol. 35, No. 10, pp. 1089-1093

Mehta Rajesh, 2000, REmoval of QRs and Impact of India's Import, EPW, Vol. 35, No. 19, May 6, pp. 1667-1671.

Rao C.H.H., 1994, Reforming Agriculture in the New Context, EPW, April 16-23, pp. 1005-1010.

Rao V.M. 1994, Agriculture and LIberalisation: Some Implications for Development Policies, EPW, April 16-23, pp. 999-1010.

3

Impact of Credit on Farming Community—A Case Study in Nellore District of Andhra Pradesh

S.K. Peeran
B. Raghavulu Naidu

Introduction

Agricultural development presupposes a greater flow of inputs as well as institutional and organisational reforms. Transformation of traditional agriculture involves an extensive use of modern technology and inputs which involve heavy financial investment. Credit is an important component in maintaining high levels of production in modern agriculture. The more highly developed the agriculture sector, the greater the amount of credit needed. In general, the demand for agricultural credit depends on the level of agricultural development of a country and its structure. Throughout the world, governments have intervened extensively in financial markets in general and rural financial markets in particular. The extent of intervention has varied from indirect measures aimed at improving the policy environment to direct steps to increase or supplement credit provided by private lenders. Many countries eager to channel funds to farmers, directed private banks to make concessional loans to agriculture or they established and supported state-owned agricultural credit institutions.

In India historically, speaking, the farmer's credit needs have been met by the local money-lender and the state, the latter playing but a subsidiary part. They were supplemented by the Co-Operative societies which were first set up in 1904 and which constitute the oldest source of institutional credit. The first survey of rural credit agencies after independence was undertaken in 1951 by the 'All India Rural Credit Survey' initiated by the Reserve Bank of India (RBI). After an exhaustive study of the rural credit situation, the committee suggested an integrated credit scheme. The recommendations of the committee accepted by the Government of India and made them an integral part of the second and subsequent five-year plans. As a consequence, there has been some quantitative progress in the institutionalisation of agricultural credit. Meanwhile the review undertaken by the All India Rural Credit Review Committee (RBI 1969) found that the co-operatives have not geared up to the expectations in mobilising deposits and disbursing credit to agriculture. The Review committee, therefore felt that the efforts of Co-operatives had to be supplemented and recommended, adoption of multi-agency approach to provide credit to rural areas with a much larger role for commercial banks. Following social control on and subsequent nationalisation of 14 major commercial banks, these banks were advised to treat agriculture as a priority sector. The flow of credit has increased to a significant level to rural areas particularly to agriculture. But surprisingly, the benefit of such increasing credit flow bypassed the relatively weaker sections of the rural community. There were added the Regional Rural Banks (RRBs) as the third constituent of the rural credit system in 1975 to cater, especially to these sections, NABARD (1982) is an apex organisation in the field of policy, planning and operational aspects so far as credit for agriculture and rural development is concerned. Since 1969, the commercial banks have evinced growing interest in financing agriculture sector. They are extending financial support to agriculture both directly and indirectly. Commercial banks have provided their finance directly to farmers in the form of short-term crop loan for agricultural production and medium and long-term loans for increasing the productivity of the land. Indirect finance to farmers is made by providing advances for the distribution of fertilisers, other inputs and also through financing Primary Agricultural Credit Societies. The banks finance the operations of Food Corporation of India and the State Governments, their agencies, for food procurement. Stocking and distribution of agricultural inputs is another sector whose needs are increasingly met by banks. The public sector banks have launched a number of scehems viz Lead Bank Scheme, Differential Rate of Interest Scheme, Integrated Rural Development Programme, Self-employment Scheme for Educated Unemployed Youth, Self-Employment Programme for Urban Poor and Service Area Approach.

Besides expansion of branch network in the rural areas and semi-urban areas, the commercial banks also opened specialised branches. These specialised branches are meant to overcome the practical difficulties relating to manpower, high cost operation and follow up of financed farmers who are scattered over a wide area. But the past experience reveals that the distribution of agricultural credit among various categories of borrowers has been unequal and inequitable. Further the effects were more pronounced in the advanced regions like deltas. The high incidence of over dues in the agricultural credit system has become a major constraint to the expansion and smooth delivery of credit. An attempt is made to examine the impact of commercial bank credit on the farming community in Nellore District of Andhra Pradesh.

Methodology

A detailed analysis relating to the impact of commercial bank credit on the farm employment, income and overdues is undertaken in this study. Nellore district has been selected purposefully since it comprises various levels of irrigation, varying soils and other natural environments. The area of the district is classified into delta and non-delta regions. The mandals of Parlapalli and Nellore rural from delta area and Duttalur and Varikuntapadu from non-delta area have been selected at random. One revenue village in each mandal has been selected by using random sampling method. Parlapalli village from Parlapalli mandal, Kakupalli from Nellore rural, M. Naidupalli from Duttalur, Varikuntapadu from Varikuntapadu mandal are selected. The cultivator households borrowing from commercial banks are identified and stratified into four strata-marginal, small, medium and large-based on the size of their ownership holding. Ten per cent of households of each strata is randomly selected. To assess the real impact of bank finance on farming community, there is a need to compare them with non-borrowers. Ten per cent of households are randomly selected from each strata. The total sample consists of 178 of which 84 borrower households and 94 non-borrower households. The profile of sample villages revealed that the delta villages is comparatively agriculturally advanced with more irrigation, whereas the non-dalta villages are backward with a low level of irigation facilities. The important crops grown in the study area are paddy, tobacco, sunflower, citrus and pulses.

Extent of Bank Credit Given to Sample Households

Due to adoption of modern technology, high yielding variety of seeds, chemical fertilizers, pesticides etc. the credit requirements off all the farmers have gone up substantially. The extent of credit borrowing particulars in sample villages from both institutional and non-institutional sources are given in Table 1.

Table 1: Borrowings from Institutional and Non-institutional Credit Agencies

(in Rs.)

		Delta Villages		*Non-delta Villages*	
	Credit agency	*Parlapalli*	*Kakupalli*	*M. Naidupalli*	*Varikuntapad*
(a)	Co-operatives	1,37,900 (17.64)	77,000 (12.96)	80,000 (9.83)	30,000 (6.62)
(b)	Commercial banks	3,26,800 (41.81)	2,61,000 (43.94)	3,09,000 (37.96)	2,00,000 (44.15)
(c)	Total institutional Credit	4,64,700 (59.45)	3,38,000 (56.90)	3,89,000 (47.79)	2,30,000 (50.77)
(c)	Non-institutional Credit	3,17,000 (40.55)	2,56,000 (43.10)	4,25,000 (52.21)	2,23,000 (49.23)
Total (c+d)		**7,81,700** **(100.00)**	**5,94,000** **(100.00)**	**8,14,000** **(100.00)**	**4,53,000** **(100.00)**

Note: Figures in brackets indicates the percentages to total credit.
Source: Field data.

Table 1, reveals that the institutional credit dominates the agricultural credit structure in the delta villages (Parlapalli with 59.45 per cent and Kakupalli 56.90 per cent) and in one non-delta village, Varikuntapadu with 50.77 per cent. In both the delta and non-delta villages, commercial banks account for a major share of 41.81 per cent in Parlapalli, 43.94 per cent in Kakupalli, 37.96 per cent in M. Naidupalli and 44.15 per cent in Varikuntapadu. The share of agriculture credit by Co-operatives is low when compared to commercial banks in both the delta and non-delta villages. It is 17.64 per cent in Parlapalli, 12.96 per cent in Kakupalli, 9.83 per cent in M. Naidupalli and 6.62 per cent in Varikuntapadu. It implies that commercial banks are playing a major role in providing agricultural credit in delta and non-delta villages.

Purpose-Wise Borrowings in Sample Villages

Purpose-wise borrowings in sample villages among different purposes is presented in Table 2.

The village-wise analysis shows that the share of crop loan is 41 per cent in Parlapalli, 46 per cent in Kakupalli (delta villages), 26 per cent in M. Naidupalli and 41 per cent in Varikuntapadu (non-delta villages). Between delta and non-delta villages, the crop loans are higher in delta villages. It may be concluded that the crop loans are the major portion of agricultural credit provided by the credit institutions in the study are followed by purchase of land, purchase of machinery, other purposes, well digging/filter points and well deepening. The borrowings for the purpose of purchase of land occupies the second place in Parlapalli (26 per cent), and purchase of machinery occupies the second place in Kakupalli (22 per cent). Similarly, other purposes like levelling the land, fencing, construction of bunds and canals etc (21 per cent) in M. Naidupalli and purchase of land in Varikuntapadu (24 per cent) occupies the second place.

Table 2: Purpose-wise Borrowings in Sample Villages

(in Rs.)

S. No.	Purpose	Delta Villages		Non-delta Villages	
		Parlapalli	Kakupalli	M. Naidupalli	Varikuntapadu
1.	Crop loans	3,21,150 (41.08)	2,71,000 (45.62)	2,14,000 (26329)	1,84,000 (40.62)
2..	Purchase of land	2,00,340 (25.63)	1,04,000 (17.51)	1,28,000 (15.72)	1,10,000 (24.28)
3.	Purchase of machinery	1,27,210 (16.27)	1,31,000 (22.05)	79,000 (9.71)	52,000 (11.48)
4.	Well digging/Filter points	41,000 (5.25)	58,200 (9.80)	1,40,000 (17.20)	57,000 (12.58)
5.	Well deepening	–	–	79,000 (9.71)	20,000 (4.42)
6.	Other purposes	92,000 (11.77)	29,800 (5.02)	1,74,000 (21.37)	30,000 (6.62)
	Total	**7,81,700 (100.00)**	**5,94,000 (100.00)**	**8,14,000 (100.00)**	**4,53,000 (100.00)**

NB: Figures in brackets indicates to total credit.
Source: Field data.

Purpose-Wise Borrowings Per Acre and Per Household

Purpose-wise borrowings per acre and per household in delta and non-delta villages is shown in Table 3.

Table 3: Purpose-wise Borrowings Per Acre and Per Household in Sample Villages

(in Rs)

Sl. No.	*Purpose*	*Delta Villages*		*Non-delta Villages*	
		Per acre	*Per household*	*Per acre*	*Per household*
1.	Crop loans	3,030	16,000	1,060	8,470
2.	Purchase of land	1,560	8,230	640	5,060
3.	Purchase of machinery	1320	6,980	350	2,790
4.	Well digging/Filter points	510	2,680	530	4,190
5.	Well deepening	–	–	260	2,110
6.	Other purposes	620	3,290	540	4,340

Source: Field data.

Table 3 indicates that the crop loans per acre and per household from institutional and non-institutional agencies account for a major part of the total loans in all the villages under study. The per acre and also per household crop loans and more in delta villages (Rs. 3,030 and Rs. 16,000) when compared to non-delta villages (Rs. 1,060 and Rs. 8,470).

The Borrowing of Crop Loans

The borrowing of crop loans per acre and per household from different credit agencies is presented in Table 4.

The Table 4 indicates that the crop loans per acre and per household from institutional agencies are more delta villages (Rs. 3,050 per acre and Rs. 16,100 per household) than in non-delta villages (Rs. 1,260 per acre and Rs. 10,040

Table 4: Amount of Borrowing of Crop Loans Per Acre and Per Household from Different Credit Agencies

(in Rs)

Sl. No	*Purpose*	*Delta Villages*		*Non-delta Villages*	
		Per acre	*Per household*	*Per acre*	*Per household*
1.	Co-Operatives	960	5,060	270	2,190
2.	Commercial Banks	2,092	11,040	990	7,850
3.	Institutional Credit (1+2)	3,050	16,100	1,260	10,040
4.	Non-institutional Credit (1+2)	1,890	9,970	1,370	10,940
5.	Total Credit (3+4)	4,940	26,070	2,630	20,980

Source: Field data

per household). In the case of average borrowing per acre and also per household from co-operatives and commercial banks are low in non-delta villages when compared to delta villages. Credit per acre is higher in delta villages (Rs. 1,890) than non-delta villages (Rs. 1,370). However, credit per household is higher in non-delta villages (Rs. 10,940) than delta villages (Rs. 9,970).

Impact of Credit on Employment

Bank finance for agricultural activities led to increase in the cropping intensity, irrigated area, and particularly under labour intensive high yielding varieties of crops. These changes resulted in employment generation of the sample households. The following Table 5 shows the generation of employment to the borrower and non-borrower sample farmers.

The Table 5 indicates that in delta villages, employment per household among borrowers was worked out to 1,380

Table 5: Employment generation among sample households (per household in person days)

Sl. No.	Category	Borrowers	Non-borrowers	Variation	%of increase
	Delta Villages				
1.	Marginal Farmers	643	593	50	8.43
2.	Small Farmers	486	424	62	14.62
3.	Medium Farmers	159	152	7	4.61
4.	Large Farmers	92	90	2	2.22
	Total	**1,380**	**1,258**	**121**	**9.61**
	Non-delta villages				
1.	Marginal Farmers	589	553	36	6.51
2.	Small Farmers	436	398	38	9.55
3.	Medium Farmers	135	126	9	7.14
4.	Large Farmers	86	82	4	4.88
	Total	**1,246**	**1,159**	**87**	**7.51**

Source: Field data

person days and while it was 1,259 person days among the non-borrowers resulting an increase of 9.61 per cent. In non-delta villages, employment among the borrowers was 1,246 person days per household and among non-borrowers, it was 1,159 person days resulting in an increase of 7.51 per cent among borrowers over non-borrowers. The category-wise employment generation shows that the small farmers occupy the first in generating more employment followed by marginal, medium and large farmers respectively in the study area. The same trend is witnessed in delta and non-delta villages. It indicates that the impact of bank finance on generation of additional employment is significant for marginal and small farmers when compared to medium and large farmers.

Impact of Credit on Farm Income

The supply of farm credit generally has positive relation with the farm income. The cheap and subsidized institutional credit further facilities the farmers to enhance their farm income., In order to ascertain the monetary benefits of all investment, it is necessary to estimate farm income. The farm income is the net income from the cultivation of crops. The excess of gross farm income over the operating cost is regarded as farm income or net farm income. The distribution of farm income among borrowers and non-borrowers is presented in Table 6.

Table—6 Distribution of Farm Income Among Borrowers and Non-borrowers

S.No.	*Category*	*Per acre income in delta villages*		*Per acre income in non-delta villages*	
		Borrowers	*Non-borrowers*	*Borrowers*	*Non-borrowers*
1.	Marginal farmers	1,180	960	320	140
2.	Small farmers	1,630	760	130	150
3.	Medium farmers	2,100	1,350	1,160	1,190
4.	Large farmers	1,590	2,630	1,190	1,100

Source: Field data

It is observed from Table 6 that in delta villages the average farm income of borrowers per acre is higher than non-borrowers in all categories of farmers except in the case of large farmers. In delta villages, among borrowers the farm income is low in the case of large farmers because they have given their lands for lease. But they have availed credit from banks and they might have used for other purposes. In non-delta villages, per acre farm income in all categories is higher among borrowers than non-borrowers except in respect of small farmers.

Table 7: Farm Income of Borrowers and Non-borrowers

(in Rs)

Item	*Delta Villages*	*Non-delta villages*	*Variation %*
Per household income of borrowers	34,460	22,320	35.23
Per household income of non-borrowers	27,020	22,860	15.39

Source: Field data.

To measure the effect of institutional credit on per household farm income, the comparison of farm income is made between borrowers and non-borrowers in delta and non-delta villages. Per household farm income is presented in Table 7.

From the Table 7 it is noticed that the average farm income per household is higher among borrowers than the non-borrowers in delta villages. In non-delta villages, the farm income per household is higher among the non-borrowers than the borrowers.

Repayment Performance

Repayment of the loan together with interest thereon by the borrowers is central to the smooth functioning of institutional credit. The incidence of overdues in the agricultural credit system has been increasing over the years and has been steadily eroding the financial soundness of the system. This, therefore, has been and continues to be the burning issue. Mounting overdues have adverse effects on both the institutions and farmers and very often lead to the collapse of the institutions. An attempt is made to analyse the repayment performance among borrowers

Table 8: Distribution of Total Demand, Repayment and Overdues in Different Categories of Sample Farmers

Sl. No.	*Category*	*Total demand*	*Payment*	*% of repayment*	*Overdues*	*% of overdues*
1	MF	1,73,180	50,230	29	1,22,950	71
2	SF	1,42,380	50,730	36	91,650	64
3	Med F	3,13,460	1,24,450	40	1,89,010	60
4	4F	3,51,330	1,56,580	45	1,94,750	55
	Total	**9,80,350**	**3,81,990**	**39**	**5,98,360**	**61**

Source: Field data.

financed for agriculture to augment income and employment in the study area. The distribution of total demand, repayment and overdues in different categories of sample farmers is presented in Table 8.

Table 8 reveals that out of the total amount of Rs. 9,80,350 due from 84 borrowers, Rs. 3,81,990 were paid which accounts for 39 per cent. The repayment performance is nor satisfactory since only 39 per cent of the amount was repaid by the sample borrowers. It is clear that there is a significant variation between the category of farmers (Marginal, Small, Medium and Large) with regard to total demand, recovery and overdues. Further the table indicates that the repayment trend was higher in the case of large and medium farmers (45 per cent and 40 per cent respectively) than the small (36 per cent) and marginal farmers (29 per cent). Overdues of credit is high in respect of marginal farmers in the study area (71 per cent) followed by small farmers (64 per cent), medium farmers (60 per cent) and large farmers (55 per cent).

The recovery performance varied considerably among the borrowers of delta and non-delta villages. Repayment performance of the sample farmers in delta and non-delta villages is presented in Table 9.

It can be observed from the Table 9 that the repayment performance among the borrowers in delta villages is 42 per cent whereas in non-delta villages, it is 36 per cent. The repayment performance in non-delta villages is not satisfactory since only 36 per cent of the total amount was repaid by the sample borrowers.

Farm size-wise analysis indicates that the repayment among marginal and small farmers is lower when compared to medium and large farmers respectively in delta and non-delta villages. The overall percentage of recovery is higher in delta villages than non-delta villages. Overdues among marginal and small farmers are higher in non-delta villages (72 per cent and 67 per cent respectively) and lower in

Table 9: Category-wise Repayment Performance of Sample Borrowers in Delta and Non-delta Villages

Sl. No.	*Category*	*Total Demand*	*Re-payment*	*% of Recovery*	*Overdues*	*% of Overdues*
				Delta Villages		
1	MF	58,800	18,360	31	40,440	69
2	SF	88,760	33,190	37	55,570	63
3	Med. F	1,61,000	27,210	42	93,790	58
4	LF	2,41,060	1,00,300	47	1,13,760	53
	Total	**5,22,620**	**2,19,060**	**42**	**3,03,560**	**58**
				Non-Delta Villages		
1	MF	114,380	31,870	28	82,510	72
2	SF	53,620	17,540	33	36,080	67
3	Med. F	1,52,640	57,240	38	95,220	62
4	LF	1,37,270	56,280	41	80,990	59
	Total	**4,57,730**	**1,62,93036**	**2,94,800**	**64**	

Source: Field data.

delta village (69 per cent and 63 per cent respectively). The lower percentage of overdues is noticed in the case of large farmers in both delta (53 per cent) and non-delta villages (59 per cent). Thus, the sample farmers in delta villages are comparatively better equipped in repayment of loans than their counterparts in non-delta villages.

On the basis of findings of the study it would be useful to make some suggestions that would help to improve the recovery of agricultural loans:

1. In our field study, from discussions, it would found that there was dissatisfaction among the borrowers who have repaid fully. The honest borrower who made repayment have reasons to feel cheated because they could not avail the benefit of debt relief for having paid the loan on time during 1990's. Hence the customers who observe credit discipline and regularly repaying the loan should receive a recognition by quoting like lower rates of interest and at times giving some mementos in the recovery camps.
2. These were a few borrowers who stated that there are some people who are habituated not to repay. Hence, there should be support from the Government and the Government may consider possibility of coming out with laws to punish heavily the wilful defaulters who some times join hands not to repay the bank loans.
3. Some of the borrowers mentioned that they were not reminded of repayment. A publicity campaign to educate farmers about the need for repayment of banks' dues should be launched by the Government through newspapers, radio/television network as is being done in the case of family planning, health and conservation of energy etc.
4. The existing laws are hardly conducive to the recovery of bank dues. The realisation of banks

dues' through the court, besides being lengthy in procedure, and is quite expensive too. Hence, special tribunals should be set up for recovery of agricultural loans. Narasimham Committee has also shown full agreement with recommendations of Tiwari Committee for setting up of special tribunals which could expedite the recovery process.

5. The legal process being cumbersome, branch managers of banks be vested with powers for auction without intervention of the courts whereby the recovery process will be speeded up at a low cost.

6. The wilful default on the part of borrowers must be treated as an act of criminal offence and criminal proceedings may be initiated after a definite period, say 3 years.

7. A common cause of overdues in the study area in the incidence of cyclones and floods in delta areas and drought in non-delta areas which result in heavy damages to the standing crops and result in heavy default in repayment of loans. An appropriate solution, it is felt, would be the crop insurance. The present crop insurance scheme which has been in operation since 1985, suffers from several short-comings. Firstly, the scheme is confined to a few crops viz, paddy, wheat, millets, oil seeds and pulses in certain defined areas notified by the State Government. Secondly, State Government has been selecting only such crops and areas where crop yields are unstable, with the result that the cross-subsidisation has not been possible in the study area, because some of the crops which were grown not covered under the existing crop insurance scheme. Hence, the scheme should cover and principal crops cultivated and should be compulsory at least to all borrowing farmers.

4

Impact of WTO on Indian Agriculture

D. Harinarayana,
L.D. Sudhakara Babu,
and K. Ravi Babu

I

Agriculture, both in India and Andhra Pradesh, is undergoing through a crisis situation. Agricultural sector realized less than on per cent growth in the last two years. Public investment in agriculture declined during 80s and 90s. Stocks of food grains worth 50m tons piled up with Food Corporation of India. Agricultural prices declined. Farmers have been resorting to distress sales or maintaining stocks unsold with them. Added to this, Chinese rice may compete in Indian market. This appears to be the case in Andhra Pradesh also in respect of most of the crops. What could be the causative factors that have let to such a situation? Are they international and/or domestic factors? Among international factors, WTO is considered to be major constraint on Indian agriculture.

Against this background, an attempt is made in this paper to analyze the role and impact of WTO on agriculture in India.

II

WTO emerged out of GATT against the background of significant changes in international economic environment. The role of trade in achieving growth is viewed differently. Hither to held controversy relating to the choice of trade strategy between import substitution and export orientation is no longer controvertible. The decade of 1990s witnessed waning protectionism and increasing liberalization in world trade. Globalization led to increased competition based on the integration of the world economy in terms of prices and costs. State regulation of the economy gave way to liberalisation. The role of the state is to be confined to that of a promoter and the other roles as regulator, planner and producer are to be phased out over a period of time. Hence public sector as a dominant producer is to give way to private sector. Therefore, exports through trade liberalization are expected to be leading economic growth rather than import substitution through protection. Against this changed international economic environment, World Trade Organisation (WTO) was established in 1995 mainly to create markets across and establish competitive markets. It has designed a set of rules to be followed by member countries in formulating their own trade policies.

III

The important concerns relating to agriculture as contained under the rules of WTO are: market access, domestic support to producers, export competition and patenting of seeds.

In relation to market access, any member of WTO should allow imports of agricultural goods atleast a minimum of 3 per cent of consumption requirements which would be increased to 5 per cent by the year 2004. Tariff barriers have to be reduced and non-tariff restrictions will have to be eliminated. Transparency is to be realized in trade. This is made possible with only tariff restrictions, that too at reduced levels. In India, quantitative restrictions are being eased out and tariff levels are reduced. Allowing imports of food items into India worsens Indian agricultural situation. Fine variety of Chinese rice is already made available in some parts of India at Rs. 7/- a kg. This adds to the existing excess supply situation, which may further depress prices. Cost of production in agriculture appears to be on the increase. Reduced profits to farmers would reduce investment in agriculture. Already public investment in agriculture is on the decline. Private investments though marginally increased in the 1990s have not helped in increasing over all investments. It is already established that land reforms, particularly ceilings on land holdings, prompted big farmers to divert surpluses from agriculture to urban sector or industry.

The second concern of WTO relates to domestic support measures, which by 2004 have to be reduced. In India, market price support has been the main stabilizing mechanism of food grain prices, particularly in states like Punjab. The price as fixed by the Government has, to a significant extent, set the market price for crops like wheat and paddy. If the support is reduced by 2004, price instability could set in the producer and the farmers may be severely affected. Unstable agriculture always crippled the economy as evident from the past experience.

The third concern relates to ensuring export competition. Reduced protection to agriculture in the developed countries, where subsidies range from 20 per cent to 30 per cent, provides increased scope for agricultural exports from developing countries. Ultimate elimination of tariff and non-tariff barriers on agricultural imports into developed countries provides price advantage to developing countries. India can take advantage in enhancing agricultural exports. However, India has to compete with other developing countries. The age-old contention of Chenary and others that developing countries will have to compete for the limited market in these goods may still appear to be relevant even today. Further, trade liberalization in agricultural goods is slower than trade liberalization in manufactured goods. There also exists scope under WTO to restrict agricultural imports into developed countries under anti dumping legislation. Even then, it is evident that the share of India relating to agricultural goods in world exports have been declining. The World Bank projected, on the eve of the establishment of WTO, that India will have advantage in agriculture related goods. Processed food items were identified as some of the dynamic items. Historically, export policy governing agricultural exports has mostly been negative under the assumption that these exports mostly face inelastic world demand with respect to price. But Indian share has been eroded due to the operation of elasticity of substitution.

Fourthly, Trade Related Intellectual Property Rights (TRIPS) as applicable to agricultural concern with patenting of seeds. 'Patent' is of two types–product patent and process patent. While product patent is followed in US, Europe and other countries, process patent is followed in India. Patenting refers to granting of monopoly to the innovator of any new product over production and marketing on a commercial scale of that product. In the field of agriculture, patenting of seeds is the major area of controversy. Firstly, as implied, seed patenting is supposed to restrain Indian farmers from using seeds, which they retain out of output to be used as seed. They also have the practice of selling surplus seeds to fellow farmers in that or neighbouring villages. The farmers cannot commercially sell seeds. However, with the establishment of a Plant Breeders' Rights (PBR) system in the near future, provision will be made for 'farmer's privilege' whereby "a farmer is completely free to use the farm-saved seed of a protected variety for growing subsequent crops either on his own land or on leased-in land or for traditional exchange in the village community. As long as the farmer continues to be a grain producer and does not convert himself into a 'commercial seed seller' of the protected variety, he would be unaffected by the PBR"[3]. It is known that Indian seed industry is basically cereal-based. It is estimated that in the most developed states of India, self retained seeds share about 80 per cent in the case of wheat, 62 per cent of paddy, 55 per cent of cotton, 59 per cent of rapeseed and 91 per cent of mustard/potato out of the total requirement of seeds[4]. TRIPS/PBR system in such a case may not affect the existing practice. Moreover, world export market in seeds is at present mainly concentrated around hybrid corn, sorghum, fodder crops, soybean and vegetables. In India, the use of HYV seeds and thereby green revolution has also generated controversy. On the negative side, adoption of green revolution is responsible for the "neglect of traditional farming practices and local cultivars...... Farmers from Kolar district in Karnataka...... cultivate HYVs of rice for sale in the market while for domestic consumption they grow traditional varieties, as these are believed to be more nutritious."[5] Further HVYs could not cope with annual floods in Assam. One of the local varieties called 'Boka' can be consumed directly by soaking in water, without any cooking. This implies that technology should be adopted to sustain local life. Moreover, use of high dose of fertilizers, pesticides under Green revolution package might endanger local environment. These arguments may not be tenable from the viewpoint food needs of the economy consequent to high population growth. Ultimately in the long-run the seed market potential is going to be high in India. Indian Agricultural research institutes comprising agricultural Universities, ICAR and other institutes have been involved in the development of hybrid sees. "India has the distinction of developing the first cotton and pearl millet hybrids in the world."[6] This also applies to horticulture and plantation crops. There exists a significant gap between requirement of certified quality seeds and its present distribution which varies from 28 per cent in the case of rapeseed and mustard to 50 per cent in the case of cotton and paddy and 75 per cent in the case of wheat, in1992-93. It is here that domestic institutes and MNCs could compete with each other in the invention of new seed varieties.

Another question that might crop up is whether prices of these seeds, particularly from MNCs, could be so high as to exploit the farmers. As could be observed, hybrid variety of seeds are more productive[7], which could influence the farmer's decision to go for these seeds. Another view held is that the WTO regime provides an opportunity to Indian farmers to get latest seeds from any part of the world. This could be true only in the long-run, if more attention is paid by international scientists on the invention of new seeds that confirm to the cropping pattern in India.

IV

It may be inferred that the present provisions of WTO in terms of market access and domestic support are not conducive from the view point of prospects of growth. Seed patenting may not negatively impact Indian agriculture in the near future. Liberal trade in agricultural goods may not benefit India significantly given its cost/price constrains. Added to this, the current scenario of reduced public investment, negligible incentives for private investment and free or liberal food imports worsens growth prospects of Indian or Andhra Pradesh agriculture. What is true of India is also mostly relevant to Andhra Pradesh agriculture.

REFERENCES

1. Sidhu, M.S. "Impact of Intellectual property Rights on the Indian Seed Industry". *Indian Journal of Agricultural Economics,* Vol. 54, No. 3, July-September, 1999. p. 372.

2. Ibid., page 372.
3. Gadgil, M. and Utkarsh, G. "Intellectual Property Rights and Agricultural Technology : Linking the micro and the macro-scales", *Indian Journal of Agricultural Economics,* July-September, 1999, p. 338.
4. Sidhu, M.S. Op. Cit., p. 374.
5. Sidhu, M.S. as quoted in Op. Cit., p. 377.

5

Challenges to Agriculture in India Under W.T.O. Regime

R.G. Desai

Introduction

The developing countries have had the least decisive role in the General Agreement on Tariffs and Trade (GATT)/ World Trade Organization (WTO); but their weakness may be much more damaging now than before because of the newly emerging features. Firstly, the WTO is increasingly spreading its coverage to new areas; secondly, the impact of agreements of the WTO (as compared to GATT) and their operation is much wider and deeper for the economies of the developing countries; and thirdly, the economies of the developing countries are less stable than before with the fear of exposure to the uncertain external environment. Against this background, the developing countries are expected to strengthen their capacity with respect to the negotiations in the WTO. In some developing countries, particularly in India, we have noticed a number of debates on various issues related to WTO which are encouraging but the level of preparedness falls short of requirement.

The recent emerging challenges related to the WTO add strains to the developing countries in expanding their export prospects. Firstly, the subjects and patterns of negotiations have now become much more complex than in the past. Secondly, the role of the developing countries in the WTO negotiations has gone over a significant change. Thirdly, the developed countries have started taking up these negotiations with a new determination to expand the access of their economic entities in the developing countries. Fourthly, the developed countries, particularly the major ones, are more coordinated in their objectives and methods in the WTO, whereas the developing countries have still to demonstrate their solidarity.

In this paper these challenges related to WTO are discussed and possible solution are recommended for policy-makers.

Major Points Emerging Under Agreement on Agriculture (AOA)

1. The AOA covers five major areas under WTO namely: Market Access, Domestic Support and Exports Competition, Trade Related Investment Measures and Trade Related Intellectual Property Rights.
2. Non-tariff barriers such as Quantitative Restrictions are to be replaced by tariffs to provide the same level of protection.
3. Developed countries need to reduce their tariffs by an average of 36% over a period of 6 years and developing countries to comply similarly by 24% over a period of 10 years.
4. Minimum level of access for imports of agricultural products (as a share of domestic consumption) should not be less than 3%. This minimum level is to rise to 5% by year 2000 for developed countries and 2004 for developing countries.
5. Developed countries have to reduce their Aggregate Measure of Support (AMS) by 20% over

6 years and Developing countries have to comply similarly by 13% over a period of 10 years. Least developed countries are exempted from this.

6. The agreement distinguishes between types of subsidies and defines three kinds of subsidies viz., Prohibited, Actionable and Non-actionable subsidies.

   (i) Prohibited subsidies: These cover the subsidies having direct influence on trade.

   (ii) Actionable subsidies: Include those subsidies having adverse effects on the trade prospects of the other signatories.

   (iii) Non-actionable subsidies: These can be either product specific or non-specific subsidies involving assistance to industrial research and pre-competitive development subsidies. These also include subsidies provided to disadvantaged regions to participate in the main stream of development.

7. Green Box and Blue Box measures provide subsidies to be excluded from AMS.

8. Peace provisions aim at the reduction of the likely hood of disputes or challenges on Agricultural subsidies.

9. Special Safeguard provisions provide safeguards against sudden price troughs in domestic market and surges in imports affecting the interest of farmers.

10. Agreement on Sanitary and Phyto Sanitary Measures provide ensuring the regulations given under the agreed rules.

11. Agreement on Trade Related Aspects of Investment Measures (TRIMS). This seeks to ensure that countries do not invest in order to create artificial

advantage and affect the trade interest of other countries (Article III of GATT).

12. Agreement on Anti-Dumping. No country will be allowed to dump the produce in the country of destination by creating an artificial price differential.
13. Agreement on Customs Valuation. This gives customs administration right to request further information on the imports wherever they have doubt about the accuracy as well as the declared value of the imported goods.
14. Agreement on pre-shipment inspection. A review process is allowed under this agreement for the inspection of the goods prior to the shipment.
15. Agreement on Safeguards. This allows the member countries take safeguard actions against any specific incidence in order to protect the domestic industry from serious effects. Safeguard measures would not be applicable to a product from a developing country member, if the share of developing country in the imports of the product does not exceed 3%. This will be done if the developing country having less than 3% import share collectively accounting for not more than 9% of the total imports of the product concerned.
16. Agreement on Trade Related Intellectual Property Rights (TRIP) recognize that there are varying standards of protection and enforcement of IPRs. It is necessary that there is standardization in the regulations across the countries.

We have following important aspects to face:

- Providing the minimum market access of 3% in the domestic market.
- Tariffication and removal of Quantitative Restrictions. Use of bounded tariff rates.

- Aggregate measure of support less than 13.33%
- Actionable subsidies to be removed
- Sanitary and Phyto sanitary standardization
- Trade in services and human capital
- Standardization of Trade Related Intellectual Property Rights Regime
- Ensuring environmental safeguards and protecting against entry of welfare deteriorating technologies.

Implications at National Level

1. As regards the Aggregate Measure of Support (AMS) we do not have immediately worry about reducing subsidies as India is far from the required threshold of AMS. It is on the contrary necessary to take advantages under 'Green Box' in order to direct the subsides to the Sunrise sectors as well as creating infrastructure for facilitating trade. It is also possible that subsidies can be directed to the backward regions under Resource Poor Region Clause from where the exportable surplus can be harnessed.

2. India should argue that the AMS can be recomputed by deflating it with an appropriate infrastructure availability index. This will facilitate an appropriate comparison between two countries having large gap in the infrastructure development.

3. It is necessary now to pay serious attention to the tariff rates and leverages provided under the tariff policy to regulate the trade sector. This should be attempted specifically for the commodities that will involve the welfare of domestic producers. India should take advantage of the bounded tariffs provided under the agreement.

4. India being high population density country, the market share of 3% or 5% of the base level of 1986-88 would be quite negligible at present. Therefore, it is not correct to assume that the import surges would be possible even under the de-regulation of quantitative restrictions. In addition to this India should take advantage of the special safeguard measures provided for arresting import surges in the predominant traditional consumption items. It will be an urgent task to create a monitoring mechanism for observing the market share of imports.
5. India can consider providing the non-actionable export subsidies in terms of reduction in export marketing costs and national and international transportation charges.
6. It is necessary that the sanitary and phyto sanitary measures as applicable in the countries of destination are made known to the exporters and a dialogue on suitably incorporating these into such requirements by the country should begin. India should insist on a common Codex suitable for our products.
7. India should lobby and argue for protection of the environmental safeguards and against the entry of welfare deteriorating technologies it is necessary to think about a measure which indicates welfare deteriorating technologies. It is necessary to think about a measure which indicates welfare deterioration index similar to the type evolved for Aggregate Measure of Support.
8. At the country level, it is necessary to have a common platform of the representatives from different states to continuously monitor the international market situation. AMS, Safeguard Measures, Sanitary and phyto sanitary measures,

anti dumping, Technical barriers to Trade, share in domestic, markets and considerations of incentives in the trade sector.

Suggestions

In order to further the interest of agriculture in the national perspective and that of the country at the international level, the following proposals are suggested:

1. Participation of the State in Policy Discussions at the Centre: Agriculture being a State subject it is necessary that the Central government discusses withe the State regarding the policies pertaining to the Sector. Similarly state government should prepare a policy with full participation from district levels.
2. Request for funds under EXIM Policy: In the exim policy modifications declared by the Ministry of Commerce on 1st April 2000 urged the States to participate vigorously in order to improve the export performance. A scheme has been evolved for granting assistance to the States on the basis of their export performance for development of export related infrastructure. To begin with, an allocation of Rs. 250 crores is proposed for the current year which would be suitably increased in the subsequent years. The amount would be utilized by the States for complementary export related infrastructure, such as roads connecting the production centers with Ports, research and development of state specific ethnic products, development of cold chains for agro exports, development of minor ports. The state government should take advantage of this.
3. Domestic Price Advantage: Government of India must take into account the domestic cost of

production of crops and agriculture commodities while fixing the import tariffs.

4. Green Box Measures: Government of India should work on the Environmental fragile zone concept in order to derive benefits under WTO provisions.
5. Environmental Safeguards: Government of India through the WTO Cell and in collaboration with the forest, horticulture, ground water board and environmental activists should create a charter of requirements to be argued with WTO while ensuring the environmental safeguards. It is feared that the international trade competition may have serious detrimental effect on bio-diversity, import of welfare deteriorating technologies, serious environmental hazards and neglect of basic requirement for the quality of life.

Specific Suggestions

1. Some of the support measures like crop insurance, drought proofing expenditure etc should be considered. These will have welfare augmenting effect on the sector and may not inflate the AMS.
2. It is essential to establish a Nodal agency at the State level to deal with the emerging situations and to continuously monitor the State's role in international trade.
3. The State agency should ensure horizontal integration across departments within the state and vertical integration with the central agency.
4. It is essential to intensify investment in the agricultural sector both at the Centre and the State level. The private component of such investment should be more effective to augment the growth performance.
5. High value specialized crop sector should be concentrated upon in order to encourage export orientation.

6. The spice derivative industry has been facing non-tariff barriers under sanitary and phyto sanitary measures. It is essential to argue the case of these commodities effectively.

7. Indian coffee is also likely to face the situation of orchratoxin contamination. It is necessary to work on both economic and technological fronts to confront the problem.

8. Seed industry in some states is emerging the recent past. It was felt that agro processing and seed industry need full encouragement to participate in the world market.

9. Investment in research and development has not been up to the mark. It is necessary to step up R and D investment to competie with the efficient production processes.

REFERENCES

1. Dr. Prem Nath, Interim Report on WTO and Related Issues, Dept. of Agriculture and Horticulture, Govt. of Karnataka, Bangalore 2000.

2. Gulati, Professor Ashok, *"Indian Agriculture in an Open Economy"* In 'India's Economic Reforms and Development' Essays for Manmohan Singh OUP 1998.

3. Srinivasan, Professore T. N. *"Developing Countries and the Multilateral Trading System"* West View Press, Harper Collins 1998.

4. Gulati, Professor Ashok *"Economic Times,* Bangalore September 1999, p. 10.

6

The Calamitous Agreement on Agriculture (AOA) Need for a Fresh Look

Abdul Noor Basha,
V. Badari Naravana Raju

Lower agricultural prices are due to WTO mandated lower import tariffs, which came upon us while the Government slept or was otherwise busy demolishing a mosque or building a temple. Even though Indian rice, milk vegetables, fruits and flowers are the cheapest in the world, our agriculture lacks the necessary packaging and transport infrastructure to take advantage of the double-edged WTO and export these products. Instead, we are today drinking Australian Orange Juice and eating New Zealand Apples, while our own fruits are not in the field for lack of marketability.

—Professor Subramanian Swamy
The Hindu, dated February 24, 2001

The frightening prediction has come true though at a much speedier pace than apprehended. A growing volume of evidence during the past couple of years apparently suggests that the Indian farmer has been hit by an unprecedented crash in crop prices. The farmers are now gripped by frustration and unrest as never before during the post-independent ear.

In the context of Andhra Pradesh, the ongoing spell of price crash has extended even to the food crops including paddy, pulses and groundnut, besides several commercial crops like cotton, tobacco and chillies, which were traditionally popular for a series of suicides by debt and price hit farmers.

One cannot afford to forget that all this devastating crisis in the farm sector has emerged only after the Indian Government signing in indecent haste, the widely opposed WTO accord in April 1994. Dr. M.S. Swaminathan, India's visionary farm scientist and institution builder has bursted in agony against the spirit of WTO in one of this recent interviews and remarked that the WTO's implementation in the last six years has shown that countries such as India have been unfairly treated with Industrialized nations not adhering to the WTO norms. He categorically stated that our farm families are experiencing all the negative impact of the WTO agreement, mostly owing to our inaction or lack of timely action against the western bias[1].

The Draconian AOA

One of the most hazardous arguments of WTO singed in Marrakech (April 1994) is related to bringing agriculture under the jurisdiction of the WTO, with the formulation of the Agreement on Agriculture (AOA), with three disastrous clauses namely, Market Access (Tariffication), Domestic Support and Export Competition. That the intent of these clauses of AOA was basically undemocratic, biased against TWCs has sufficiently been proved in the Indian context by now.

The section dealing with the Market Access is designed to provide market for the western countries, mostly the US, for agricultural products especially for food grains. Two aspects are involved in it, first is the conversion of Quantitative Restrictions (QRS) into 'Tarrfication', while the second is the reduction of import duties. Obviously, both of the measures prove harmful for the welfare of the domestic economy.

The unjust AOA legislates to provide unrestricted import of food grains by a member country to the amount equivalent to 3 per cent of the domestic consumption during a particular year. India, for instance is food surplus country, hence needs no imports. India now has an unspent food reserve of 45 million tonnes. The FCI has already declared its inability over fresh procurement of food grains from the price-hit farmers. Amidst these critical circumstances, present the market access clause of AOA, only a revisionary tendency in the domestic market.

Consequent to trrafication, a reduction in the base tariff under a time bound programme-by 24 per cent over 10-years in the case of developing countries and by 36 per cent over six years for developed countries was envisaged.

Domestic support is designed to protect the industrial countries against competition from the developing countries through withdrawal of subsidies. The export competition section deals with the direct subsidisation of agriculture and enjoins all countries to reduce direct and monetary subsidies to exporters in agriculture.

Unequal 'Free Trade" Source for the Calamity

Several eminent intellectuals characterized the WTO agreement as 'unequal and unjust trade bargain' for Indian agriculture, Vis-a-vis the western countries, on the following issues:

1. Import Tariffs

While India is forced to cut drastically its import tariffs on a variety of goods, the industrialized countries have been restoring to heavy tariffs on imports to protect their markets. A few examples of heavy imports by the industrialized countries are given here under:

Country	*Product*	*Import tariff (%)*
U.S.	Sugar	244
	Groundnut	173
E.U.	Beef	213
	Wheat	167
	Mutton	144
Japan	Wheat	352
	Wheat based goods	388
	Rice	200
Canada	Butter	360
	Cheese	281
	Eggs	236

The average tariffs in the OECD countries for 1995 stood at 214 per cent for wheat, 197 per cent for barley and 154 per cent for maize, while the average import tariffs in he developing countries were 94 per cent for wheat and 90 per cent for maize and rice.

As against this, India brought down its tariffs from as high as 250 per cent to just 15 per cent allowing the farmer to face the price-crash.

2. *State Support*

Similarly, while India has been withdrawing all sorts of state support to its farmers, the industrialized countries are offering mounting subsidies to their small number of farmers to help them to dump the TWCs. For example, a farmer in the OECD countries enjoys a handsome subsidy of $29,000 (Rs. 12 lakhs plus) a year. According to the recent estimates, the total support rendered to the agricultural sector by the countries comprising the OECD increased from $308 during 1986-88 to $347 bn in the three year period 1997-2000. Similarly, the US domestic support for its farmers was $25.5 bn in 1996, while in the EU it was $85 bn for their products. India's domestic support to its hundreds of millions of farmers on the other hand, worked out to a negative $23.7 bn in 1995-96, even after providing for fertilizer, electricity, irrigation and seed subsidies[2]. So, as Utsa Patnaik, distinguished Professor of Economic, has rightly remarked the terms of trade under WTO have shifted very significantly against developing countries[3].

3. Export Subsidies

Export subsidies further distort prices in the developing countries. Only 24 of the WTO's 135 members have a right to subsidies exports (as only they declared such subsidies during the Uruguay round) while only two or three agribusiness corporations (mainly from the US and EU) account for most export subsidies. Three exporters account for 93 per cent of subsidized wheat exports, two for 80 per cent of subsidized beef exports and two for 94 per cent of subsidized butter export[4].

Privileges to US–EU

To protect themselves, the industrialized countries have introduced a number of safeguards such as the 'Blue box' and 'Green box' in the AOA to provide them with cover for providing 'domestic support' for their farmer. The provisions in the Green box include policies that provide services or benefits to the agriculture or rural community, stock holding for good security, domestic food aid, investment subsidies and agricultural input subsidies for low income and reduce poor families. The provisions in the Blue box include direct payments to farmers under production limiting programmes. These subsidies are branded 'production neutral' and are excluded from the state subsidies, thus, enabling the US and EU to come out into international market with low priced products.

For example, in the US, the farmers are paid not for producing a crop, though the payment is not called the subsidies. This implies that direct payments can be provided to support incomes of farmers and that the overall cost of production will not be reflected in the price of commodities. On the other hand as Dr. Vandana Shiva argues, since income of TWC farmers are derived from production and trade and not from direct income support from governments, the farmers will be totally vulnerable to changes in global trade patterns and international prices[5].

Quite surprisingly India's ruling class has been in a great a hurry in exhibiting undue obedience to WTO directions. As a result of a dispute initiated by US against India, our Government agreed early last year itself to phase out all quantitative restrictions on imports on an accelerated schedule, much earlier to the WTO–stipulated deadline of the year 2003. The principal agricultural commodities like rice and milk powder are already committed at zero import duty. For wheat, the roller flour mills are allowed to import at zero import duty. Even the pulses are also now being imported under OGL at zero import duty. Edible oils, most of which are bound at 300 per cent import duty are now open for imports at 15 per cent duty[6]. Even in the recent Central Budget, the Finance Minister could not afford to push up the scale beyond 8 per cent. Commenting on the irresponsive position assumed by India to a reduction of import tariffs, Arun Ghosh, distinguished economist and former member of planning commission had to state that 'what is really baffling in this connection is the reduction of maximum level of tariff from 40 per cent to 35 per cent just a month prior to the announcement of the Exim policy in March 2000[7].

Of the 2,714 products that were subject to import restriction till 1999, no fewer than 1285 were shifted to be OGL in April that year. The following year, an other 715 products were decontrolled, leaving only the remaining 714 for the final phase out of import decontrol in this year. Most of these products are related either to agriculture or to small scale industries which form the basis of continued employment for teeming millions in India.

Dr. Swaminathan, therefore rightly emphasized that when the WTO agreement is reviewed this year, it is essential that we press for a 'livelihood box' that would allow us to impose QRS for an other 10-15 years and stop dumping[8].

Impact on Andhra Pradesh Agriculture

The WTO accord has been blamed by several state governments for the present price crash in agriculture. The Chief Ministers of Punjab, Haryana, Karnataka, Bihar and Andhra Pradesh have already explained their anguish against the centre's illegitimate coalition with WTO and demanded for a radical review of the international trade arrangements.

The Chief Minister of Andhra Pradesh Mr. Chandrababu Naidu had suggested during a national debate, one the implications of WTO regime that indiscriminate import of rice, edible oils and other produce and inadequate procurement by the FCI were responsible for the present crisis.

Food crops are badly affected in Andhra Pradesh. The paddy of common variety could not get more than Rs. 400/- per quintal as against the support prices of Rs. 510. Fine and superfine varieties too, were not accepted on the ground of their being damaged lot. In fact, the minimum support price itself was not commensurate with production costs. For example, the total cost of cultivation of some variety of paddy was worked out at Rs. 12.600/- per acre including land rent' of Rs. 3.000 in most of the places of Rayalaseema and Telangana regions. (even as the per acre rent touches Rs. 10,000 in coastal districts). If the production is taken at 26 quintals (about 35 bags of 75 kgs, each), the total revenue totalled only Rs. 13, 390/- @ Rs. 480/- per quintal, thus leaving a margin of less than Rs. 800/-. This little margin evaporates if interest on investment is also reckoned with. The problem of price was compounded with rice millers holding 15 lakhs tonnes and formed about 10 lakh tons out of last years production. This led for glut in the market. During 1993-94, total imports of food grains and other agricultural products that entered into the country were valued at Rs. 1,024 crores. The figure rose to high of Rs. 96,000 in 1998-99. Most of this bill included only food grains, Oil seeds, oil cakes and edible oils are by far the largest group of agricultural imports in the products freed form QRS in 1999 and 2000. The government policy of keeping edible oil prices low contributed to this phenomenon. However, there is evidence that the imports have fallen in this category after the hike in import duties in November 2000. During 1999-2000, Rs. 8,425 crores worth of oil seeds, cakes and oils were imported, while the figure has come down to Rs. 3,367 crores during 2000-2001 (April-September). These imports made a significant impact on AP farmers especially relating to groundnut.

It is quite surprising as the national daily 'The Hindu'[9] surveyed that though India is one of the largest exporters of spices, their imports into India touched Rs. 68 crores during 1999-2000, though it is even more surprising that fruits and vegetables including watermelons, cherries, raspberries, pineapples etc., were imported to the tune of Rs. 150 crores during a short span of 6 months (April-September 2000-01). The Andhra Pradesh agriculturists naturally had to face a price crisis in the QR era.

The government of Andhra Pradesh, however has been unconcerned to these adverse developments and harsh in the release of subsidies. The amount of subsidies given to agriculture and its allied activities declined sharply from

Rs. 79 crores in 1992-93 to Rs. 50 crores in 1994-95 and further to as low as Rs. 27.6 crores in 1997-98. In other words, the share of agricultural subsidies in the total subsidies affected by the government of Andhra Pradesh has fallen from 20.5 per cent in 1992-93 to just 3.94 per cent in 1997-98[10].

On the other hand, rice exports from Andhra Pradesh declined considerably. The state used to export 20 lakh tonnes of rice to other countries besides 25 lakh tonnes to other states. But in the last two years, exports declined as the state's rice was found to be more expensive at $5 to $10 a quintal. In the international market fine rice is available at a very moderate price of Rs. 750/- only. Therefore FCI or STC or any other agency is not in a position to procure and promote exports from the state. To sum up, the WTO regime, while the industrialized countries are extending the state support to its farmers, the developing countries deny even minimum assistance and hence the glut and price crash. The state and the centre in collusion with industrialized countries strongly wish to inject the myopic spirit of 'comparative advantage', thereby gradually compelling the exodus of paddy farmers into other commercial crops required by the west.

The member countries in the WTO have to step up support to agriculture and to continue all the required protective measures for domestic farmers. Unfortunately, both the centre and the state governments are moving to their MNC oriented new agricultural policies. For example, in its recently released strategy paper on agriculture, the Andhra Pradesh government is silent on several crucial issues like the resource mobilization, appointment of staff, credit to farmers and marketing outlets. Andhra Pradesh agriculture is under staffed and credit starved as commented by India's outstanding journal, The India Today. The total land under cultivation in Andhra Pradesh is 106.16 lakhs hectares, while in Karnataka and Tamil

Nadu, it is 103.90 lakhs hectares and 59 lakhs hectares respectively. But while Andhra Pradesh appointed only 4,381 farm employees, Karnataka and Tamil Nadu appointed 7,958 and 7, 680 respecrtively[12]. Similarly, out of the total Rs. 36,000 crores credit needs, the Andhra Pradesh farmers were thrown in distress to depend on the money lenders to the tune of Rs. 30.000 crores.

Though the pro WTO intelligentia argues that the WTO in itself is not a monster they tend to admit the undue influence of the rich nations by their violating the rules of the game in their favour without providing any level playing ground for this. Perhaps, with the same perception in mind, the NDA government at the centre has now been contemplating the need to raise such questions in the next AOA review. The Union Agriculture Minister Mr. Nitish Kumar had to admit recently, responding to the Congress led farmers rally in Delhi on February 25, that the ministery was examining measures for safeguards against unfavourable provision in the WTO, which would be raised during AOA review. Quite interestingly he further stated that "more particularly, it is being examined whether we can go beyond the bound rate of tariff on imports, if its impacts the livelihood of farmers. The other measure is to impose restrictions on quantity even after April 1, when India is obliged to remove QRs" 13. Even if these belated expressions of our rulers are translated into action, it would come as a miracle of Alladin's wonderful lamp for the farmers lying on the stretches.

REFERENCES

1. For a detailed interview with Dr. Swaminathan, please see the Frontline, February 16, 2001.
2. Sukumar Muralidharan "On the World Trade Front", Frontline, February 2, 2001, p. 12.
3. Thomos, Binu. S "Seattle, Will developing countries hold their own", Business line, November 29, 1999.

4. Patnaik, Utsa 'The Perils of unregulated volatility' Frontline, February 2001, p. 21.

5. Thomos Binu. S., Op. Cit.

6. Vandana Shiva, "WTO and Free Trade", CPI Publications, January 2000, pp. 15-16.

7. Gurate, Ashok; *et. al.*, "From Marakesh to Seattle–Indian Agriculture in a Globalising World", Economic and Political Weekly, October 9, 1999, pp. 29-33.

8. Ghosh, Arun, "EXIM Policy, Dreams and the Reality" Mainstream, April 8, 2000, p. 3.

9. Swaminathan Interview Op. Cit. p. 115.

10. The Hindu dated February 22, 2001, BS-1.

11. Compiled from various budget papers and Directorate of Economics and Statistics.

12. India Today, Telugu edition, February 27, 2001.

13. The Hindu dated February 25, 2001.

7

Pattern of Changes in Distribution of Workers, Landuse and Productivity in Agriculture Sector an Inter and Intra Regional Comparison in A.P.

B. Chakravarty

As a consequence of the cumulative spread affects of pre and post green revolution initiatives, the total food production in India has been in the order of 200-220 million tons for the last one decade. Now we are in a position to maintain a buffer stock to the extent of 45 million tonnes. Thanks to farming community for enabling Government of India to launch a noval scheme "Antyodaya Anna Yojana" (AAY) through which a family living below poverty line would be provided 25 kgs of food grains at a cost of Rs. 2/- per kg. of wheat and Rs. 3/- per kg. of rice. The Govt. of Andhra Pradesh has also implemented various innovative schemes for the development of agriculture and allied sectors. Some of the commendable initiatives of Government of A.P. would include—issue of Pattadar pass books and credit cards to farmers, formation of Water Users' Associations, Watershed Management Committees, Rythu Bazars, rain water harvesting structures under

Janmabhoomi with focus on "Neeru-Meeru" and water and soil conservation activities under watershed development projects.

Table 1: Plan Expenditure by Heads of Development in Andhra Pradesh from 1980-81 and 1983-84

(Rupees in Crores)

Head of Development	*1980-81 Expen-diture*	*%*	*1983-84 Expen-diture*	*%*
Ag. And Community Dev.	54.83	11.7	96.11	12.5
Irrigation and Power				
(a) Major and medium	136.09	29.1	146.29	19.1
(b) Power	136.47	29.2	141.01	18.4
Industrial and Mineral Development	21.85	4.7	67.09	8.7
Transport and Communication	38.02	8.2	51.07	6.7
Social services	78.06	16.7	260.40	33.9
Miscellaneous	1.87	0.4	5.12	0.7
Total	**467.19**	**100.0**	**767.09**	**100.0**

Source: 1. Bureau of Economics and Statistics, Govt. of Andhra Pradesh, 1980-81, pp. 168-169.

2. Bureau of Economics and Statistics, Govt. of Andhra Pradesh 1983-84, pp. 180-181

While appreciating the relentless efforts of farmers for achieving food security in India, the rural economy in general and agriculture sector in particular have been experiencing since 1990, the following features of recession:

- Increasing pressure on land resources due to alarming population growth.
- Recurrence of droughts and floods on a regular interval.

Table 2: Plan Expenditure by Heads of Development in Andhra Pradesh from 1995-96 to 1998-99

(Rupees in Crores)

Head of Development	*1995-96*		*1996-97*		*1997-098 (R)*		*1998-99*	
Ag. and Allied services	17.19	(2.7)	109.55	(3.6)	178.53	(4.8)	204.96	(4.1)
Rural Development	196.12	(6.8)	326.12	(10.7)	195.15	(5.3)	388.62	(7.8)
Spl. Area Dev.	119.78	(4.2)	36.35	(1.2)	282.08	(7.6)	493.16	(9.9)
Irrigation and Flood Area control	652.50	(22.7)	689.51	(22.6)	795.75	(21.5)	931.49	(18.7)
Energy	842.05	(29.3)	827.19	(27.1)	858.65	(23.2)	797.46	(16.0)
Industry and Mines	70.56	(2.5)	41.16	(1.3)	65.74	(1.8)	92.60	(1.9)
Transport	352.29	(12.3)	365.52	(12.0)	376.33	(10.1)	539.78	(10.9)
Science, Tech. and Even.	0.52	(-)	0.67	(-)	0.32	(-)	6.84	(0.1)
General Economic Services	3.35	(0.1)	4.76	(0.2)	39.49	(1.0)	67.58	(1.4)
Social Services	536.44	(18.7)	635.12	(20.8)	893.96	(24.1)	1425.08	(28.7)
General Services	18.58	(0.7)	16.31	(0.5)	21.24	(0.6)	24.40	(0.5)
Total	**2869.38**	**(100.0)**	**3052.26**	**(100.0)**	**3707.24**	**(100.0)**	**4971.97**	**(100.00)**

Source: Bureau of Economic & Statistics, Govt. of A.P.

- Substantial cut on the subsidies to the farm in-puts due to conditions imposed by World Bank and WTO.
- Lack of assured marketing with minimum support price for principal crops.
- Gradual decline in agriculture sector's contribution to GDP, decline in the percentage of budgetary allocations to agriculture sector.
- Suicidal deaths among farming community due to crop failures and heavy rural indebtedness.
- Growing rural urban migration due to under and unemployment in rural areas.

In the light of the above mentioned, one gets an impression that agriculture sector can no longer boast itself to be the backbone of development in India and AP is no exception. Against this backdrop, an attempt has been made in this paper to highlight the pattern of changes taken place in work force, land use and productivity of some of the principal crops during the past one and a half decade, and their implications in the rural economy.

Part—I of this paper seeks to highlight the trends of development in different spheres of economy of Andhra Pradesh and its relative position in all India on a set of important indicators of development with special reference to agriculture sector during the recent past. Part-II provides a comparative picture of pattern of change in distribution of workers during 1981-91 in Coastal Andhra, Rayalaseema and Telangana regions of A.P. on one side and changes in land utilization during 1987-88 and 1998-99 on the other. Part-III presents the changes in the percentage of area covered under major principal crops and also per hectare productivity for major crops during 1987-88 and 1998-99. The last Part deals with the summary and conclusions. Since the analysis has been confined to a few parameters of agriculture development, the conclusions drawn may not

Table 3: Gross State Domestic Product (GSDP) of AP by Industry at Constant (1993-94) Prices

Industry	*1993-94*		*1994-95*		*1996-97*		*1997-98*		1998-99	
Agriculture	17227	(29.6)	16794	(27.3)	19033	(27.5)	15472	(22.7)	19776	(26.2)
Forestry and Logging	669	(1.1)	673	(1.1)	681	(1.0)	667	(1.0)	720	(1.0)
Fisheries	1043	(1.8)	1083	(1.8)	1238	(1.8)	1226	(1.8)	1374	(1.8)
Mining Quarrying	1323	(2.3)	1425	(2.3)	1569	(2.3)	1615	(3.4)	1607	(2.1)
Primary Sector	20262	(34.8)	19975	(32.5)	22521	(32.6)	18980	(27.9)	23477	(31.1)
Secondary	12504	(21.5)	14525	(23.6)	15929	(23.0)	17026	(25.1)	17820	(23.6)
Tertiary	25461	(43.7)	26992	(43.9)	30734	(44.64)	32009	(47.0)	34233	(45.3)
Total GSDP	**58227**	**(100.0)**	**61491**	**(100.0)**	**69188**	**(100.0)**	**68015**	**(100.0)**	**75530**	**(100.0)**

Source: Bureau of Economics & Statistics, Govt. of A.P.

fully reflect the comprehensive picture of agriculture development in A.P. However, it provides some insights into the core issues of agriculture development which have substantial bearing on livelihood pattern of rural people. The major part of the data presented in this paper has been obtained from Statistical Abstracts of Andhra Pradesh for 1988, 1992 and 1999 Season and Crop Report of AP for 1998-99 and Statatical Hand book of AP for 1999-2000.

PART I

Development Profile of AP in Brief

Andhra Pradesh with 73% of its population living in rural areas in one of the agriculturally progressive states in India. As per 1991 Census, it ranks fifth with 6.65 crores population. The population growth rate over the previous census has shown a significant increase of +16.65 in 1961 and it has increased @+20.90 in 1971, +23.10 in 1981 and +24.20 in 1991. As regards gross cropped area, it ranks fifth with 1,27,83,000 hectaters during 1994-95, while in terms of net irrigated area (41.23 lakh hectares during 1995-96) it ranks fourth, claiming a share of 7.70% in the total net irrigated area of 53508 lakh hectares for all India.

The number of workers in Agriculture sector has increased from 128.23 lakhs in 1961 to 195.16 lakhs in 1991, even though there was a sharp decline in 1971 (126.25 lakhs).

The general literacy rate as per the 1991 Census is 44.1 is much below the national average of 52.2. The rural female literacy is as low as 23.72 as against the national average of 30.6.

As on 31st March 1998, as many as 18,24,689 irrigation pump sets have been energized in AP claiming a share of 15.40% in the total of 11,847,191 for all India, and as such it has taken second position after Maharashtra (21,51,191). The number as on 31st March 2000 has gone up to 19,18,712.

Table 4: Growth of State Economy

Item	*Unit*	*1997-98*	*1998-99*	*1999-2000*
Population	Lakhs	734.33	742.63	751.00
Gross Irrigation	Lakh	51.58	60.92	57.45
Production of foodgrains	Lakh tons	108.22	149.05	136.97
Production of oilseeds	Lakh tons	12.14	24.49	13.61
Production of cotton	(000) bales	13.20	15.22	15.79
Sugarcane	170 kg unit	13.46	17.87	19.67
Electricity generated	Million units	26265	27024	29632
Electricity installed capacity	Mega watts	5612	5612	5627
Enrolment of children in Primary school (I-V)	Lakhs	83.7	87.9	91.1

Source: Handbook of Statistics, Directorate of Economics & Statistics, A.P. 1999-2000, p. 2.

The population covered by one Commercial Bank (as on 31st March 1999) is 14,80,000 as against 15,07,000 for all India and it ranks 14th amongst states. In terms of per capital advances it ranks 8th with Rs. 3,485.12 and it is less than the all India figure of Rs. 3,968.44. The total regional rural bank branches in A.P. are 1121 (as on 31.3.1999) claiming a share of 7.74% in the total branches of 14,475 for all India and as such it ranks 4th.

In terms of consumption of chemical fertilizer per hectare (in kgs) during 1996-97, it has maintained 2nd rank with 134.4 kgs as against 76.00 kgs for all India. Its consumption has declined to 132.6 kgs in 1997-98 and got relegated to 4th place (provisional), but its consumption is higher than the national figure of 86.1 kgs.

The per capita consumption of electricity including utilities and non-utilities (in K.wh.) in A.P. during 1996-97 is 332 and it is less than the national average of 349. During 1997-98 also it has maintained the same trend (provisional) but its rank declined from 8th during 1995-96 to 11th (368 K.wh.) during the above two reference periods.

In terms of per capital income it has registered an increase from Rs. 7494 in 1993-94 to Rs. 13584 in 1998-99. These figures are below the national average of 9698 and 14683 for the above reference years respectively.

In terms of area under total food grains its share to all India is 5.7% (rank 7th), while in terms of food grains its share is 14,395,000 tons constituting a share of 7.1% in the total production of 202.54 million tons in India.

While the plan expenditure on agriculture and community development has been showing an increasing trend in terms of absolute figures, its percentage has declined from 11.7 in 1980-81 to 4.1 in 1998-99. Similar trend could be seen in the case of irrigation and power.

The per cent contribution of Agriculture sector to Gross State Domestic Product (GSDP) at constant (1993-94) prices has declined from 34.8 in 1993-94 to 31.1 in 1998-99, while the percentage contribution of secondary and tertiary sectors has maintained an upward trend during the same period.

PART II

Distribution of Workers During 1981-91

Trends of change/shift could be seen in different categories of workers during 1981-91 from the data furnished in Table 5. The percentage of cultivators to total population of the state has shown significant decline from 1981 (13.8%) to 1991 (11.86). Similar trend could be seen in all the three regions (Coastal, Rayalaseema and Telangana) of A.P. However within the Rayalaseema region, Kurnool district has shown a marginal increase during the above period while in Telangana region, Ranga Reddy district has shown the highest percentage shift and Warangal has shown the lowest shift.

In the case of Agriculture labourers the percentage has increased by about 2 per cent in A.P. and same trend could be found in all the three regions. As regards workers in the household sector and other workers, while the former category has shown decline, the latter category has shown an increasing trend. In both the categories the percentage shift is about one per cent across the regions and districts.

In the case of marginal workers category a declining trend could be observed in all the districts except Khammam. The percentage of shift in the non workers category is not significant except Vizianagaram and East Godavari districts.

Distribution of workers among Scheduled Castes and Scheduled Tribes could be seen for 1991 in Table 6. The percentage of Agricultural labourers among scheduled tribes is higher than that of scheduled castes in all the three regions and the highest percentage is recorded in Telangana region. However the figures are much higher in the districts of Khammam, Warangal, Nalgonda, Adilabad, Visakhapatnam and East Godavari, where concentration of the tribal population is reported to be higher.

Table 5: Distribution of Workers Engaged in Agriculture Sector: District-wise 1981 and 1991 (General)

Sl. No.	*District*	*% of Cultivators*		*% of Labour*		*% of workers in the household workers*		*% of other workers*		*% of marginal workers*		*% of non-workers*	
		1981	**1991**	**1981**	**1991**	**1981**	**1991**	**1981**	**1991**	**1981**	**1991**	**1981**	**1991**
1	**2**	**3**	**4**	**5**	**6**	**7**	**8**	**9**	**10**	**11**	**12**	**13**	**14**
1.	Srikakulam	18.0	14.38	14.8	18.24	2.2	1.66	8.4	9.71	5.6	4.65	50.9	51.36
2.	Vizianagaram	19.1	17.46	13.3	15.55	1.8	1.53	9.3	10.95	5.0	4.84	51.5	49.67
3.	Visakha	16.2	14.53	8.2	9.45	1.6	1.06	13.2	14.99	4.3	3.08	56.5	56.89
4.	E. Godavari	8.2	5.48	16.3	18.55	1.8	1.20	11.7	11.57	2.6	1.68	59.4	61.52
5.	W. Godavari	8.9	6.25	20.5	22.92	1.5	0.92	10.4	11.44	2.1	1.82	56.7	56.65
6.	Krishna	8.2	6.58	17.5	20.33	1.3	1.00	13.0	13.92	3.1	1.47	56.9	56.70
7.	Guntur	11.2	9.36	19.6	23.45	1.3	0.55	11.6	12.29	3.7	2.11	52.6	52.14
8.	Prakasam	13.9	12.22	19.4	23.49	2.1	1.16	9.2	9.32	4.2	2.45	51.3	51.36
9.	Nellore	11.6	9.11	18.7	20.43	1.8	1.27	10.6	12.04	3.7	2.77	53.6	54.38

(Contd...)

1	2	3	4	5	6	7	8	9	10	11	12	13	14
Coastal A.P.		12.1	10.6	16.7	19.16	1.7	1.15	11.1	10.80	3.6	2.76	54.9	54.51
10	Chittoor	18.4	16.21	13.5	15.04	1.5	0.99	8.8	10.74	3.6	2.95	54.2	54.07
11	Cuddapah	13.3	12.02	15.7	16.25	2.4	1.59	9.8	10.53	3.7	3.56	55.1	56.05
12.	Anantapur	17.2	15.40	15.2	16.72	1.6	1.43	8.6	9.80	3.0	2.81	54.3	53.80
13.	Kurnool	11.6	11.16	20.7	22.34	1.5	1.06	10.4	10.54	2.5	1.41	53.5	53.49
Rayalaseema		15.3	13.69	16.2	17.59	1.7	1.26	9.4	10.41	3.2	2.68	54.2	54.35
14.	Mahbubnagar	20.9	19.06	16.2	20.39	2.2	0.90	7.5	8.00	4.9	1.86	48.2	49.79
15.	Ranga Reddy	14.3	10.05	12.6	11.32	1.8	0.40	14.1	17.92	3.2	1.06	54.1	59.25
16.	Hyderabad	0.1	0.07	0.2	0.07	0.4	0.14	27.0	26.70	0.1	0.38	72.1	72.64
17.	Medak	20.3	18.50	14.3	17.05	2.2	1.30	7.9	9.47	4.7	1.46	50.5	52.22
18.	Nizamabad	18.6	16.05	13.6	15.09	5.6	6.51	10.3	10.22	2.8	1.78	49.1	50.35
19.	Adilabad	16.0	14.14	15.3	14.59	2.2	2.03	9.5	10.97	3.4	2.64	53.5	55.63
20.	Karimnagar	16.5	15.08	16.8	17.39	4.3	3.86	10.1	12.34	2.8	1.65	49.5	49.68
21.	Warangal	15.4	14.68	15.7	19.06	2.4	1.60	9.4	9.95	4.9	2.49	52.2	52.22
22.	Khammam	13.7	11.78	17.7	20.94	1.3	0.83	9.3	9.75	2.1	2.57	55.0	54.13
23.	Nalgonda	16.8	14.62	16.7	19.48	2.6	2.40	8.3	8.81	5.0	3.00	50.6	51.69
Telangana		15.2	13.40	13.8	15.53	2.5	2.00	11.4	12.41	3.5	1.89	53.8	54.77
Andhra Pradesh		13.8	11.86	15.5	17.48	2.0	1.44	10.9	11.99	3.5	2.28	54.2	54.95

Source: Census of India 1981, 1991 (for A.P.)

Table 6: Distribution of Workers Engaged in Agriculture Sector

Sl. No.	*District*	*% of Cultivators*		*% of Labour*		*% of workers in the house-hold sector*		*% of other workers*		*% of marginal workers*		*% of non-workers*	
		SC	**ST**	**SC**	**ST**	**SC**	**ST**	**SC**	**ST**	**SC**	**ST**	**SC**	**ST**
1	**2**	**3**	**4**	**5**	**6**	**7**	**8**	**9**	**10**	**11**	**12**	**13**	**14**
1.	Srilakulam	4.58	12.54	34.85	30.84	0.79	1.70	6.87	8.09	4.36	2.66	48.55	44.17
2.	Vizianagaram	4.61	20.13	31.81	27.78	0.44	1.49	9.54	3.51	4.89	5.52	48.71	41.57
3.	Visakha	1.63	39.94	19.37	8.65	1.00	0.39	16.25	2.32	2.44	6.30	59.31	42.40
4.	E. Godavari	0.99	24.71	36.73	22.23	0.28	0.99	6.20	4.09	2.91	2.37	52.89	45.61
5.	W. Godavari	1.65	12.23	42.57	31.52	0.23	1.90	5.79	8.05	1.95	1.20	47.81	45.10
6.	Krishna	2.00	4.0	40.43	31.91	0.25	2.70	6.97	10.81	1.88	1.42	48.47	49.14
7.	Guntur	2.02	4.83	4.214	36.29	0.13	1.25	6.79	9.41	2.06	2.25	46.84	45.97
8.	Prakasam	2.84	3.85	43.36	33.12	0.16	2.29	4.31	10.07	2.16	2.03	47.17	48.64
9.	Nellore	4.84	2.27	37.53	42.78	0.24	1.25	6.03	8.35	2.77	2.48	48.59	42.87

(Contd...)

1	2	3	4	5	6	7	8	9	10	11	12	13	14
Coastal A.P.		2.79	13.83	36.53	29.45	039	1.55	7.63	7.18	2.82	2.91	49.81	45.05
10	Chittoor	7.55	7.95	35.97	36.26	0.36	1.45	5.23	6.73	2.73	2.89	48.16	44.72
11	Cuddapah	3.49	5.66	37.87	27.45	0.26	6.17	5.55	7.56	3.05	4.05	49.78	49.11
12.	Anantapur	7.78	14.18	35.43	22.88	0.32	2.85	4.88	6.46	2.36	3.12	49.23	50.51
13.	Kurnool	5.33	6.11	37.23	22.42	0.43	9.55	7.04	8.56	1.22	1.52	48.75	51.84
Rayalaseema		6.03	8.47	36.62	27.25	0.34	5.00	5.67	7.32	2.34	2.89	48.98	49.04
14.	Mahbubnagar	15.54	26.35	35.89	19.18	0.36	1.54	3.45	3.22	1.69	2.03	43.07	47.68
15.	Ranga Reddy	9.19	18.99	26.00	17.29	0.14	1.16	11.73	10.17	1.24	1.32	51.70	51.07
16.	Hyderabad	0.05	0.15	0.13	0.04	0.19	0.38	28.26	30.82	0.42	0.25	70.95	68.36
17.	Medak	15.99	25.26	34.38	17.72	0.31	2.23	3.88	4.30	1.71	1.75	43.73	48.72
18.	Nizmabad	15.24	27.97	32.98	20.26	3.42	1.11	4.48	2.53	1.63	1.68	42.25	46.45
19.	Adilabad	10.67	25.38	23.51	19.86	1.15	0.67	8.34	3.02	3.26	2.96	53.07	48.11
20.	Karimnagar	11.62	20.09	36.17	25.81	1.19	2.48	5.87	6.21	1.60	1.35	43.55	44.06
21.	Warangal	7.47	28.12	36.16	20.88	0.24	0.72	6.27	2.52	2.10	3.71	47.76	44.05
22.	Khammam	3.88	20.10	35.47	26.65	0.34	0.55	7.02	3.29	2.17	3.68	51.12	45.73
23.	Nalgonda	7.87	23.31	37.65	21.03	0.41	1.52	4.65	2.40	2.96	4.00	46.46	47.74
Telangana		9.75	21.57	29.83	18.87	0.77	1.23	8.39	6.84	1.87	2.27	49.36	49.19
Andhra Pradesh		6.24	20.90	35.18	23.63	0.48	1.33	6.93	4.88	2.25	3.29	48.92	45.97

Soruce: Census of India 1981, 1991 (for A.P.)

In the case of scheduled castes, the percentage of agricultural labour is more than 1/3rd of their population while in the case of Scheduled Tribes it less than 1/4th of their population AP level. Chittoor and Nellore are the exceptions where the percentage of agricultural labourers among scheduled tribes in higher as compared to Scheduled Castes.

In the case of workers in the household sector category, though in majority of the districts the percentage is higher among scheduled tribes, the variation between SC and ST is much higher among the districts of Rayalaama.

In the other workers and marginal workers categories, the districts with predominance of tribal population have shown higher percentages among scheduled tribes while in other districts the percentages are higher among scheduled castes. Similar trend could be observed in non-workers category.

Landuse Pattern During 1987-88 and 1998-99

Changes in the land use under nine fold category for two periods i.e. 1987-88 and 1998-99 could be seen in Table 7. A declining tend could be found in all the cases except forest, land put to non-agricultural use, other follows and net area sown at A.P. level. In Krishna district, the percentage of area put to non-agricultural use is the highest (16.9%) barring Hyderabad (84.4%) during 1998-99. The land under current follows has shown a declining trend in Coastal Andhra and Telangana, while a marginal increase could be seen in Rayalaseema.

The area under net area sown is reported to be highest in Coastal Andhra (46.0%) followed by Rayalaseema (40.47%) during 1998-99.

PART III

Percentage of Area Under Principal Crops to the Total Area Sown

The trends of change in percentage of area under principal crops during 1987-88 and 1998-99 could be seen

Table 7: Landuse Pattern—Districtwise (1987-88 and 1998-99)

Sl. No.	District	Forest cultivators		Barren and unculturable		Land put of Non-Ag. Use		Culturable waste		Pasture and Grazing land		Land under Misc crops	
		87-88	98-99	87-88	98-99	87-88	98-99	87-88	98-99	87-88	98-99	87-88	98-99
1	2	3	4	5	6	7	8	9	10	11	12	13	14
1.	Srikakulam	12.0	12.1	10.7	8.9	12.7	15.3	2.0	0.08	0.4	0.1	0.6	0.4
2.	Vizianagaram	17.7	17.7	13.4	12.3	11.6	12.0	1.1	0.5	0.9	0.7	1.4	1.2
3.	Viskhapatnam	41.5	42.1	15.6	13.3	6.8	7.2	1.2	1.0	0.6	0.3	1.1	1.8
4.	E. Godavari	29.8	29.8	7.7	7.6	9.9	11.2	2.9	1.7	3.0	2.3	0.9	0.6
5.	W. Godavari	10.4	10.4	6.4	5.6	11.4	12.9	5.3	3.2	3.2	2.0	1.1	0.8
6.	Krishna	7.5	7.5	7.6	6.3	11.8	16.9	5.1	3.2	2.5	1.3	0.7	0.6
7.	Guntur	13.7	14.3	3.7	3.1	12.8	13.7	5.6	3.5	2.3	2.1	4.0	3.6
8.	Prakasam	12.3	25.8	7.6	8.8	8.4	8.5	5.4	3.9	3.5	4.0	0.7	0.6
9.	Nellore	7.5	18.7	14.9	12.2	15.1	17.6	4.7	8.0	9.1	7.6	1.2	1.3

(Contd...)

1 2	3	4	5	6	7	8	9	10	11	12	13	14
Coastal A.P.	17.2	19.82	9.7	8.67	11.0	12.81	4.0	2.78	3.2	2.26	1.3	1.21
10. Chittoor	30.2	30.1	11.2	11.0	9.3	9.5	3.5	2.8	2.8	2.4	1.7	1.7
11. Cuddapah	32.8	32.8	16.3	15.4	10.9	10.7	3.6	4.4	1.4	1.0	1.2	0.6
12. Anantapur	10.3	10.2	9.9	9.0	8.2	8.5	4.4	4.1	1.2	1.2	1.0	0.5
13. Kurnool	18.1	18.0	5.6	5.6	5.5	5.6	5.2	4.6	0.2	0.2	0.1	0.1
Rayalaseema	21.9	22.77	10.5	10.25	8.4	8.57	4.2	3.97	1.3	1.2	1.0	0.72
14. Mahbunbagar	16.4	14.4	6.3	5.2	6.0	4.2	0.8	0.6	2.4	1.3	0.3	0.3
15. Ranga Reddy	9.7	9.7	6.2	4.9	9.9	11.5	3.7	3.5	7.0	6.1	1.1	0.9
16. Hyderabad	–	–	3.4	2.0	7.5	83.4	–	13.2	0.1	–	–	–
17. Medak	7.7	10.0	6.8	6.0	6.0	7.2	2.7	2.2	4.6	2.7	0.9	0.7
18. Nizamabad	20.8	21.0	7.3	8.1	9.1	9.2	1.8	1.9	4.8	3.2	0.5	0.3
19. Adilabad	42.4	4.25	2.3	2.7	3.7	3.7	1.4	0.9	2.8	0.8	0.5	0.5
20. Karimnagar	21.2	21.0	7.1	7.0	6.0	6.7	1.7	1.6	4.8	4.6	0.6	0.6
21. Warangal	28.8	28.9	4.4	4.1	4.2	4.6	1.8	2.6	4.2	3.9	0.4	0.3
22. Khammam	47.9	47.0	6.5	5.7	4.0	7.9	1.7	1.3	3.0	1.3	0.5	1.2
23. Nalgonda	6.0	6.0	6.3	6.3	6.4	8.4	1.7	2.4	6.8	4.6	0.6	0.7
Telangana	24.1	20.05	5.7	5.2	5.9	14.68	1.7	3.02	0.6	2.85	0.6	0.55
Andhra Pradesh	21.2	22.6	8.2	7.6	8.2	9.1	3.1	2.8	0.9	2.5	0.9	0.8

S. No.	*District*	*Current follows*		*Other follows*		*Net area sworn*		*Total geographical area (in hectares)*
		87-88	**98-99**	**87-88**	**98-99**	**87-88**	**98-99**	
1	**2**	**3**	**4**	**5**	**6**	**7**	**8**	**9**
1.	Srikakulam	23.7	7.6	1.7	0.5	35.8	54.6	584290
2.	Vizianagaram	15.5	1.5	1.0	0.7	37.1	53.0	630038
3.	Visakhapatnam	10.0	2.6	1.7	0.4	21.2	31.0	1134284
4.	East Godavari	4.6	3.0	2.0	1.4	38.8	41.9	1081843
5.	West Godavari	7.1	2.8	3.0	1.1	51.8	60.8	779538
6.	Krishna	10.3	2.4	2.5	2.8	51.7	58.6	879694
7.	Guntur	3.5	1.6	2.7	1.0	51.3	56.8	1132824
8.	Prakasam	15.2	7.4	1.37	8.3	32.8	32.4	1714062
9.	Nellore	8.3	2.3	13.9	7.2	24.8	24.9	1316042

(Contd...)

1	2	3	4	5	6	7	8	9
Coastal A.P.		10.3	3.46	6.0	2.6	37.1	46.0	1028068
10.	Chittoor	2.7	4.2	9.0	6.2	29.1	31.8	1498770
11.	Cuddapah	3.2	4.4	5.0	4.2	25.0	26.1	1537838
12.	Anatapur	11.1	8.2	6.9	5.1	46.6	52.9	1913492
13.	Kurnool	5.9	7.8	7.2	6.6	51.9	51.1	1760034
Rayalaseema		6.0	6.15	7.0	5.52	39.2	40.47	1677533
14.	Mahbubnagar	17.8	24.3	4.2	6.2	45.3	43.1	1847241
15.	Ranga Reddy	14.1	13.7	7.4	8.3	40.5	41.0	753247
16.	Hyderabad	16.2	0.1	1.3	0.2	2.4	0.9	23002
17.	Medak	17.6	18.2	8.2	5.1	45.1	47.4	951903
18.	Nizamabad	22.6	17.1	1.4	7.6	31.5	31.3	806215
19.	Adilabad	7.0	3.2	2.5	11.5	37.0	33.7	1188499
20.	Karimnagar	20.1	20.6	1.5	4.3	36.7	33.3	1620381
21.	Warangal	15.3	4.4	3.3	10.1	37.3	40.8	1283552
22.	Khammam	6.9	3.7	1.3	1.4	27.8	29.1	1580936
23.	Nalgonda	22.7	20.1	5.1	11.3	43.9	39.9	1422324
Telangana		15.4	12.54	3.6	6.6	38.4	34.05	1147730
Andhra Pradesh		11.4	8.5	5.2	5.5	38.1	40.3	27440049

Source: Compiled from Statistical Abstracts—1988 and 1999, Bureau of Economics and Statistics, Govt. of A.P.

from the data furnished in Table 8. A declining trend could be seen in case of cereals and millets and pulses while in the other food category, it has reported an increase at AP level. In case of cereals and millet crops, the percentage of area has declined by 7.4% for Andhra Pradesh but its decline is more in Rayalaseema (15%) and less in Coastal Andhra (3%). Telegana has maintained an intermediate picture with 7%. The area under food crop has declined by more than 4% for A.P. Its decline is about 10% in Rayalaseema. In the total food category, Coastal Andhra has shown relatively better picture during 1998-99. The overall picture suggests that the percentage of area diverted towards non-food crops has been increasing gradually for various reasons.

Productivity of Principal Crops

The details of yield per hectare for some of the principal crops during 1987-88 and 1998-99 could be seen from Table 9. As regards rice, Coastal Andhra, has reported higher yield per hectare as compared to Rayalaseema and Telangana during the both reference periods. Except Anantapur, in all the districts, yield per hectare during 1998-99 is reported to be higher as compared to 1987-88. Within Coastal Andhra, West Godavari has reported the highest yield during both the reference periods while the districts of North Coastal Andhra, have reported relatively lower yields which are less than the stage average. In Telangana region, except Nalgonda during 1987-88 and Karimnagar during 1998-99 all the districts have reported the yields less than the stage average of 2258 kgs during 1987-88 and 2812 kgs during 1998-99. In case of Jowar, Rayalaseema has reported higher yield during both the periods followed by Telangana. As regards pulses, black gram has reported lower yield per hectare in majority of the districts of A.P. during 1998-99 as compared to 1987-88 while reverse trend could be seen in the case of green gram. Coastal Andhra and Rayalaseema have reported equal quantum (834 kgs) during 1987-88, while

Table 8: Percentage of Area under Principal Crops to the Total *Area Sown during 1987-88 and 1998-99: District-wise*

S.No.	District	Total Cereals & Millets		Total Pulses		Other Crops	Food	Total Crops	Food
		87-88	**98-99**	**87-88**	**98-99**	**87-88**	**98-99**	**87-88**	**98-99**
1	**2**	**3**	**4**	**5**	**6**	**7**	**8**	**9**	**10**
1.	Srikakulam	31.8	48.8	22.9	18.9	9.7	10.5	64.4	78.2
2.	Vizianagaram	25.3	35.1	15.5	15.2	8.9	15.1	49.7	65.4
3.	Visakhapatnam	51.0	46.7	8.6	6.3	19.0	30.9	78.6	83.9
4.	East Godavari	66.2	54.9	6.9	17.1	12.4	16.4	85.5	88.3
5.	West Godavari	72.6	67.8	3.3	1.5	14.3	20.6	90.2	90.4
6.	Krishna	55.9	56.9	23.4	16.4	10.4	16.2	89.7	85.5
7.	Guntur	36.7	38.4	22.8	20.2	6.4	8.3	65.9	66.9
8.	Prakasam	45.5	30.7	7.9	15.3	6.6	7.3	60.0	5.3
9.	Nellore	68.5	64.1	2.7	4.2	7.1	13.0	78.3	81.3

(Contd...)

1	2	3	4	5	6	7	8	9	10
Coastal A.P.		52.4	49.26	12.9	12.79	10.3	15.36	75.6	77.02
10.	Chittoor	28.1	22.8	4.3	2.7	11.7	21.3	44.1	46.8
11.	Cuddapah	37.4	19.5	3.6	5.9	11.9	15.3	52.9	40.7
12.	Anantapur	23.1	9.6	6.0	5.9	5.5	4.6	34.6	20.1
13.	Kurnool	43.8	23.8	6.3	9.2	10.7	11.1	60.8	44.1
Rayalaseema		33.4	18.92	5.5	5.92	9.4	13.07	48.3	37.92
14.	Mahbubnagar	56.1	42.9	11.1	11.2	1.4	3.4	68.6	57.5
15.	Ranga Reddy	57.7	49.1	17.6	19.8	8.6	12.6	83.9	81.5
16.	Hyderabad	45.9	42.6	-	-	22.5	18.7	68.4	61.3
17.	Medak	71.0	57.0	14.0	18.9	8.4	13.8	93.4	89.7
18.	Nizamabad	70.1	63.5	11.0	8.6	12.8	16.1	93.9	88.2
19.	Adilabad	50.4	44.0	16.2	14.8	2.7	4.2	69.3	63.0
20.	Karimnagar	58.4	62.0	21.8	8.4	5.3	7.7	85.5	78.1
21.	Warangal	53.0	42.0	17.6	12.3	2.8	7.7	73.4	62.2
22.	Khammam	49.0	45.8	24.1	13.9	9.2	14.4	82.3	74.1
23.	Nalgonda	55.0	50.9	13.6	11.0	1.1	4.5	69.7	66.4
Telangana		56.9	50.0	16.0	11.89	4.7	10.31	77.6	72.2
Andhra Pradesh		49.8	42.4	12.4	11.7	7.9	11.7	70.1	65.8

Source: Statistical Abstracts: 1988 and 1999, Bureau of Economics and Statistics, Govt. of A.P.

Table 10: Yield Per Hectare of Principal Crops-Distrctwise (1987-88 and 1998-99)

S.No.	*District*	*Rice*		*Jowar*		*Black Gram*	
		87-88	98-99	87-88	98-99	87-88	98-99
1	**2**	**3**	**4**	**5**	**6**	**7**	**8**
1.	Srikakulam	1417	2015	607	350	245	410
2.	Vizianagaram	920	2342	493	300	834	660
3.	Visakhapatnam	1079	1334	435	319	834	714
4.	East Godavari	2422	2632	443	319	834	311
5.	West Godavari	2730	3370	490	488	834	427
6.	Krishna	2265	3104	638	481	1072	763
7.	Guntur	2522	3212	311	774	738	843
8.	Prakasam	2256	3211	476	857	834	592
9.	Nellore	2608	3086	568	474	834	656

(Contd...)

1	2	3	4	5	6	7	8
Coastal A.P.		2382	2700	499	484	834	597
10.	Chittoor	2383	2595	731	577	834	777
11.	Cuddapah	2375	2819	862	864	834	669
12.	Anantapur	1970	2518	815	1505	-	667
13.	Kurnool	2140	2846	911	1058	834	676
Rayalaseemas		2243	2694	876	1001	834	697
14.	Mahbubnagar	1795	1988	582	587	435	342
15.	Ranga Reddy	2055	2207	790	490	435	649
16.	Hyderabad	2224	2765	623	680	-	-
17.	Medak	1886	2470	701	645	435	135
18.	Nizamabad	2087	2750	891	753	492	491
19.	Adilabad	936	2366	486	707	403	365
20.	Karimnagar	1431	3239	724	551	435	457
21.	Warangal	2113	2759	645	559	435	456
22.	Khammam	1692	2739	897	650	435	514
23.	Nalgonda	2393	2903	335	410	435	350
Telangana		2032	2618	621	603	435	375
Andhra Pradesh		2258	2812	662	695	771	608

Table 10: Yield Per Hectare of Principal Crops-Distrctwise (1987-88 and 1998-99)

S.No.	*District*	*Green Gram*		*Groundnut*		*Sugarcane (Gur)*		
		87-88	**98-99**	**87-88**	**98-99**	**87-88**	**98-99**	**98-99***
1	**2**	**3**	**4**	**5**	**6**	**7**	**8**	**9**
1.	Srikakulam	237	376	615	978	6747	7549	69417
2.	Vizianagaram	369	249	610	917	6299	6730	62000
3.	Visakhapatnam	354	512	1159	1256	4584	4740	43585
4	East Godavari	419	449	857	1457	7212	8216	76557
5.	West Godavari	365	413	1118	1352	8084	8564	80152
6.	Krishna	362	633	1003	1408	8265	11697	107453
7.	Guntur	374	558	1259	1550	6659	7634	70643
8.	Prakasam	356	663	902	1244	6732	7133	66566
9.	Nellore	531	269	1716	2292	6842	9656	89165

(Contd...)

1	2	3	4	5	6	7	8	9
Coastal A.P.		358	458	951	1383	6724	7991	181393
10.	Chittoor	344	614	1254	1106	7092	10575	97834
11.	Cuddapah	342	622	1031	1026	7463	11068	100579
12.	Anantapur	342	625	1023	1157	6502	10753	98816
13.	Kurnool	346	579	1017	891	7499	10477	97297
Rayalaseema		344	610	1080	1045	7095	10718	98631
14.	Mahbubnagar	490	441	602	758	7777	8350	77800
15.	Ranga Reddy	490	645	799	1167	7300	9131	83622
16.	Hyderabad	—	—	—	—	—	—	—
17.	Medak	649	438	772	1217	6982	8037	74325
18.	Nizamabad	486	582	839	2303	7872	9728	88912
19.	Adilabad	267	264	644	1279	7680	9621	87333
20.	Karimnagar	743	266	946	1121	7699	9698	87927
21.	Warangal	437	552	917	1054	—	—	—
22.	Khammam	514	480	969	1209	6807	9046	82991
23.	Nalgonda	177	507	618	785	7671	8976	82459
Telangana		487	463	772	1210	7386	9073	83171
Andhra Pradesh		460	462	997	1082	6902	8363	77231

*Data relates to Cane

Source: Bureau of Economic and Statistics, Govt. of A.P.

Rayalaseema (697 kgs) has surpassed Coastal Andhra (597 kgs) during 1998-99.

In case of Green gram, Telangana has reported higher yield (487 kgs) followed by Coastal Andhra (358 kgs) during 1987-88 where as Rayalaseema (610 kgs) has maintained its supremacy over Telangana (463 kgs) and Coastal Andhra (458 kgs) during 1998-99.

The productivity of groundnut has shown an increasing trend except in the districts of Rayalaseema during the above reference periods. But in the case of sugarcane, all the districts have reported higher yields during 1998-99 as compared to 1987-88.

Summary and Conclusions

Though there has been an increase in the overall food production, the per capita availability of food-grains has been declining due to alarming rise in the population. The decline to the extent of 7.4% in the area under cereals and millets over a period of ten years suggest that the farmers have been gradually shifting to the other lucrative crops which consume more water. This may not be a good sign in the Telangana, Rayalaseema regions where the ground water level has been depleting at a faster rate and therefore less water consuming crops are more suitable in these regions.

The area under food crops has declined be more than 4% at state level and its decline is about 10% in Rayalaseema. This trend may pose a serious threat to the food security in the years to come.

Decline in the percentage of cultivators and an increase in the agriculture labour during 1981-91 (except Hyderabad, Ranga Reddy and Adilabad where proportion of agriculture labour has declined), suggests that a significant proportion of cultivators has been converted into agriculture labourers due to non-viability of their land holdings and sale of their lands. In view of the serious under and unemployment situation in rural areas, the proportion of agricultural labourers and farmers (including small and marginal) migrating to urban areas has been on increasing side. The situation might also be attributed to the failure of self and wage employment schemes implemented over a period of time.

The proportion of marginal workers could be seen higher in Telangana region while the proportion of non-workers has shown an increasing trend.

There percentage of area covered under forest has increased from 21.2% in 1987-88 to 22.6% in 1998-99. Even the all India figure is in the same order (22.5%) during 1995-96. It takes a long way to even come closer to the globally prescribed figure of 33.3%. If this trend is allowed to continue, we may face serious drought and destification hazards in the future. Therefore the Central and State Governments should bring all the culturable waste and barren lands under green cover on a war footing with an adequate budgetary support.

There should be consistent and comprehensive agriculture policy at state level which can instill a feel of security and confidence among the farming community. This policy should also envisage legal provisions/safeguards such as protection of lands belong to SC/ST and other backward communities, comprehensive insurance coverage for all crops in case of droughts and floods, assured supply of power at affordable rates with preferential treatment to the backward areas.

The population growth should be checked with incentives to the poor, formulation of sustainable self-employment schemes with adequate unit cost for the deserving under the unemployment youth may bring some positive changes in the rural economy in the near future.

8

Agriculture and Environment

K.B. Reddy

Introduction

Population explosion, increase in per capita consumption of food, decrease in per capita cultivable land, outbreak of insect pests, frequent crop failures in spite of rapid increase in subsidies in the form of productive and protective inputs have been responsible for food shortages. Even as the yields of many of the cultivated varieties are showing signs of decline, we are required to improve our food production. In order to meet the growing demands for food, fodder, fibre etc., agriculture has undergone revolutionary changes starting from the sustainable food gathering agriculture to the present day hyper-intensive agriculture with serious environmental consequences.

Odum (1972) defined an agroecosystem as a solar powered, human subsidised, man made ecosystem. It cannot sustain itself without the frequent intervention of man. Agriculture ecosystems are ecologically unstable and they are maintained by the subsidies provided in the form of productive inputs such as fertilizers, irrigation, ploughing, weeding etc., and protective inputs like pesticides.

Among the three things i.e. air, water and food required by every living organism, plants alone have the capacity to prepare food materials from the simple inorganic materials of CO_2 and water. All other organisms are either directly of indirectly dependent on plants for their food. Plants play a very important role in environmental protection but agriculture is highly harmful to the environment. The impacts of agriculture on environment have been discussed in the paper under report. The main objective of the paper is to show how environment gets damaged by agriculture to all those who are very much concerned about food production as well as environmental protection.

Environmental Impacts of Agriculture

The impacts can be classified into the beneficial and harmful effects. Except for production of food, fodder, fibre etc., and generation of direct rural employment and indirect urban employment, agriculture is not beneficial to the environment. Agriculture can not even play the traditional role of a tree which is capable of storing carban dioxide for several years as most of the agricultural production is either consumed or degraded or incinerated within about a year. But the adverse and harmful effects of agriculture on environment are far and wide. These effects are divided into direct and indirect categories.

Direct Effects of Agriculture on Environment

They are due to the (1) pesticides (2) fertilizers (3) irrigation (4) ploughing and (5) monocultures.

Pesticides

Pesticides are used to protect the crops from insects, fungi, bacteria and other predators and pathogens. Pesticides like DDT, HCH, Aldrin, Dieldrin etc. have been responsible for major ecological problems such as food chain concentration (also known as biomagnification and bioamplification), toxicity to the non-target organisms; persistence and accumulation in food, fodder, milk, meat etc. leading to adverse health impacts and elimination of several species of predator birds; insect and pest resistance to pesticides; loss of balance of nature owing to elimination of beneficial organisms; pollution of water, contamination of food, acute toxicity to the pesticide sprayers, decline in productivity and increase in food production costs.

There are over 400 chemical factories in India manufacturing 55 basic pesticides. A total of 134 pesticides have been registered in India. The use of pesticides is growing at the rate of 2 to 5% in different States. Yet the average pesticide consumption of 300 g/ha is far below the U.S average of 1kg/ha and that of 10 kg/ha in Japan (Dhaliwal *et a1.*, 1996). But the average DDT level of 28 ppm in human body tissues in India is much higher than 2.01 in Netherlands, 2.3 in Germany, 3.0 in UK, 3.1 in Denmark, 10.1 in Itally, 12.4 in Hungary, 13.4 in Poland and 18.1 in Israel. This is mainly due to improper use and handling of pesticides in India. An analysis of cereals, pulses, vegetables, oils, meat and eggs obtained from different states in India showed that 12 to 96% of the samples contained upto 60 ppm of DDT and HCH (Kalra and Chawla, 1981). More than 62% of honey samples contained DDT, HCH, lindane, aldrin and heptachlor. Vegetable oils, milk, butter are also found to contain very high levels of organochlorine pesticides. The average daily intake of DDT by infants in India through milk was estimated to be 45 μg/kg/day as against the permitted limit of 5 μg/kg/day. Most popular brands of butter in India contained upto 6.33 ppm of DDT and 4.34 ppm of HCH as against the tolerable limit of 1.25 ppm. In contrast, only 5 out of 1141 samples of dairy products in USA contained DDT residues upto 2 ppm. Similarly, the pesticide residues in European dairy products were negligible. (Singh and Dhaliwal, 1994). Human milk was found to be contaminated with pesticides beyond the safety limit.

Results of the dietary intake of DDT and HCH residues in India showed an average daily intake (ADI) of 231 and 128.5 ug per person respectively, The ADI of DDT in india is much higher when compared to that of 6.5 in US and 12.0 ppm in UK (Singh 1982). Rachel Carson (1962) clearly demonstrated the involvement of DDT in biological magnification and elimination of aquatic birds in her famous book, 'Silent Spring'. Subsequently, DDT was banned in USA in 1972 but allowed their companies to manufacture DDT for export purposes. India restricted the use of DDT in 1992 but DDT residues continue to increase in biological systems due to their persistence. Pesticides have been responsible for loss of balance of nature and natural regulation of pests owing to destruction of beneficial insects and the resistance of harmful insects to insecticides.

Chemical Fertilizers

More than 50 million tonnes of Nitrogenous fertilizers and 13 million tonnes of phosphatic fertilizers are used globally in agriculture in addition to the 2 million tonnes of phosphorus used in detergents. The current level of chemical fertilizer consumption in India is 18 million tonnes. Chemical fertilizer usage in India stands at 70 kg/ha, as compared to 236.5 in china, 422.4 in South Korea, 432.7 in Japan and 687.1 kg/ha in Neatherlands. The nutrient consumption in India has increased steeply between 1960 and 1983 (Dhaliwal and Singh, 1994). Replacement of organic fertilizers by chemical fertilizers in India has been responsible for hardening of soils, decrease in per cent pore space, increase in rate of run off and soil erosion. Enrichment of water bodies with nutrients lost along with flood waters has been responsible for eutrophication of water. Increase of nitrate content in drinking water has been responsible for nitrate toxicity. The nitrate content of ground water in U.P., Haryana and Punjab was found to range between 240 and 1310 mg/L as against the WHO recommended safe limit of 45 mg/L (Handa 1987). High concentrations of nitrate in drinking water causes methamoglobinaemia (blue baby syndrome) in infants and the formation of carcinogenic N-nitrosamines and nitroamides in adults.

Agriculture, particularly, cultivation of paddy has also been responsible for production of methane, a serious green house gas. Extensive cutlivation of only a few high yielding varieties of crops has been responsible for loss of tremendous amount of crop biodiversity. For instance, before green revolution more than 50,000 varieties of paddy were in cultivation in India but now there are only 5 or 6 varieties in cultivation.

Agriculture was intended to provide food to the people. Paradoxically, every jump in food production was followed by a similar increase in population resulting in greater shortage of food. Man is described as the pest of nature and agriculture has helped to promote the growth of this pest which caused extensive damage to the environment threatening his own survival. Agriculture is also responsible for production of large amount of biodegradable waste which is burnt in open. Incineration of biomass causes air pollution due to the production of CO_2, CO, oxides nitrogen, BAP (Benzapyrene), Polycyclic aromatic hydrocarbons (PAH) and suspended particulate matter (SPM).

Irrigation

In order to sustain crop production against the vagaries of monsoon, assured irrigation is necessary. Impounding of flowing waters, construction of huge reservoirs have been responsible for water logging, salinity, toxicity, silting, submergence of large areas, dislocation of people, reservoir-induced earthquakes, apart from huge investment which in most cases is not cost-effective.

Ploughing is responsible for rapid oxidation of organic carbon present in soils and the release of CO_2. It is also responsible for soil erosion, loss of nutrients and the consequent eutrophication of water bodies.

Indirect Effects of Agriculture on Environment

Fertilizer factories and pesticide plants besides consuming fossil fuels and electricity, are also responsible for pollution of air, water and soil. The world's worst industrial accident which claimed thousands of lives in Bhopal was caused by a pesticide plant. Leakage of ammonia in a number of fertilizer plants is a routine event. Since the fertilizers and pesticides are used in agriculture, the environmental effects of these industries are attributable to agriculture.

Transportation of food, fodder, fibre over longer distance, from the place of production to the ultimate consumer requires lots of fossil fuels and road space to move the vehicles. This vehicular pollution and some of the road accidents are thus traceable to agriculture. Live-stock and dairy, which are dependent on agriculture production have their own impacts on environment. Many of the food processing industries which make use of agricultural produce cause pollution of air, water and soil besides generating solid waste.

Thus agriculture is harmful to the environment but the survival of man is linked to agriculture. Hence agriculture can not be abandoned for the sake of environment. Therefore, there is a real need to develop the technologies, processes and practices which are eco-friendly so that agriculture becomes sustainable. Integrated pest management practices (IPM), integrated water and nutrient management, integrated soil management and integrated crop management practices involving the use of biopesticides, biofertilizers, pest and drought resistant crops, crop rotation, polycultures in stead of monocultures, use of biodegradable pest—specific pesticides are some of the techniques currently under experimentation. There is also a need to breed and select the plants in order to ensure that they survive and produce in the changing environment. Global warming is feared to disrupt monsoon cycles and pest cycles. A slight rise in earth's average temperature by 1° C is likely to reduce production of cereals in tropical countries like India by 10%. Thus agriculture depends on environment and environment is threatened by the agriculture. The need of the hour is to balance these two so that they go together.

REFERENCES

Dhaliwal, G.S. and B. Singh, 1994. Agrochemicals in Indian Agriculture, Retrospect and Prospect. In: G.S. Dhaliwal and B.D. Kansal (Eds.) Management of Agricultural Pollution in India. Commonwealth Publishers, New Delhi, pp. 1-21.

Dhaliwal, G.S., Sanga, G.S. and P.K. Ralhan, 1996. Fundamentals of Environmental Sciences, Kalyani Publishers, Ludhiana, p. 374.

Handa, B.K., 1987. Nitrate content of ground water in India. Fertilizer News 32 : 6,

Kalra, R.L., and R.P. Chawla 1981. Impact of pesticidal pollution in the environment. Journal of Bombay Natural History Society 78 : 1-15.

Odum, E.P., 1972. Ecosystem theory in relation to man. In : J. Wiens (ed.). Ecosystems: Structure and function. Oregan State University Press.

Rachel Carson, 1962. Silent Spring. Haughton Mifflin, Boston.

Singh, P.P., 1982. Insecticide residues in food in Punjab. A typical dietary survey. Ph.D. thesis, Punjab Agriculture University, Ludhiana.

Singh, B. and G.S. Dhaliwal, 1994. Pollution hazards of insecticides in agriculture. In : G.S Dhaliwal and B.D. Kansal (Eds.) Management of Agricultural Pollution in India. Commonwealth Psublishers, New Delhi, pp. 22-56.

9

Watershed Development—A Study of Naravagadda Watershed

C. Ratnam
B. Chinna Rao

A large part of Indian agriculture is operated under rainfed conditions and the benefits of Green Revolution did not percolate to it. The rainfed crops are mostly the low value crops with high risk of crop failure and their yields are very low and unstable. So the people depending on it are poor. Also the rainfed agriculture is causing acological degradation in terms of soil erosion resulting from the cultivation of marginal and sub-marginal lands, deforestation, over grazing etc. On the other-hand the yields of irrigated crops have reached almost a stagnation stage. Therefore, for meeting the increasing demand for food, achieving sustainable agricultural development and alleviating the rural poverty, there is a need to promote technology suitable for dry farming. For promoting such technology, various efforts are made by CRIDA, ICRISAT and Agricultural Universities in the country. One of the outcomes of such efforts is the evolution of Watershed Development Approach, a technique which replaced the high cost dry land agriculture by low cost rainfed agriculture with a greater emphasis on natural resource conservation measures besides other supportive infrastructure.

A watershed is defined as a geo-hydrological unit or a piece of land that drains out into a common point from its ridge to the outlet. The extent of the watershed depends on the nature of common drain point i.e. stream, rivulet, tributary or river. The productivity in the watershed depends on the interaction between soil, rainfall, drainage and the vegetative cover. The lack of desired water and soil conservation measures, inadequate vegetative cover are causing soil erosion, and reducing yield capabilities. To check this phenomenon and integrated approach for development of dryland agriculture through watershed development programmes is necessary.

Encouraged by the success of the watershed development programmes in the VIIIth Plan Period the Department of Agriculture and Co-operation launched the "National Watershed Development Project for Rainfed Areas" (NWDPPA) in 1991. This programme was implemented in all community development blocks where less than 308 of the arable land is under assured means of irrigation. It is a massive programme for development of rainfed agriculture with low cost technology, involving the local farmers for maintenance of soil end moisture conservation structures. It has a provision of subsidised cost to reap the benefits. It aims to extend the Green Revolution to the areas untouched so far by adopting farming systems approach on scientific watershed management principles involving animal husbandary, horticultural, agroforestry, fisheries, seri-culture, silvi-culture, etc.

The ultimate aim of this programme is to develop natural resource base, sustain its productivity, improve the standard of living of the rural poor and landless labourers and endesvour for restoration of ecological balance.

More **fically this programme is aimed at:

1. Conservation upgradation and utilization of natural resources like land, water, plant, animal resources in a harmonious and integrated manner.
2. Improving production environment and restoration of ecological balance through scientific management of land and rain water.
3. Reducing the inequalities between irrigated and rainfed areas and also reducing large scale migration from rural areas to urban areas.
4. Increasing employment and marketable surplus of dairy production and agriculture especially by growing cash cross like vegetables, horticulture through improved cropping systems.

The major components of this programmes are:

1. Treatment measures for soil and moisture conservation.
2. Crop production.
3. Ecological development.
4. Livestock and animal husbandary development.

The important activities under conservation measures are:

1. Contour bunding and stone terracing
2. Vegetative filter strips
3. Contour vegetative hedges
4. Gully control measures which include check dams, brushwood dams etc.
5. Contour cultivation
6. Other measures like small dugout ponds and run-off management in dugout ponds.

The production system include the activities like:

1. Crop demonstration for single crop and multiple crops.

2. Agro-forestry
3. Dry land horticulture
4. Homestead gardens/Kitchen gardens.

The unique feature of this project is greater people participation right from the start of the project till the completion of the project. In designing the project due weightage is also to low cost and easily adoptable technologies, farmers active participation in planning and their concensus in developmental activities of the project and their needs and requirements.

Naravagadda Micro Watershed in Visakhapatnam district, Andhra Pradesh was selected as one of the places for the implementation of NWDPRA scheme during the VIII plan period. This Micro Watershed has 13 villages spread over a total gregraphical area of 7417 hectares, of which only 6041 hectares are actually falling in the watershed. The total arable area with in the watershed is 4067 hectares of which the area irrigated is 1310 hectares, leaving an area of 2757 hectares for various soil and moisture conservation treatment. Also 974 hectares of non-arable land in the watershed is available for various treatment measures.

In this watershed area the scheme was implemented in a phased manner during the period 1991-92 to 1996-97 with an estimated cost of Rs. 98,159 lakhs. The soil and moisture conservation measures taken up were contour vegetative hedges, contour graded bunds and trenches with vegetative support, repairs to the existing structures, brushwood structures gully stabilization, repairs and maintenance of existing run-off management structures, field drains, contour cultivation across the slopes, contour furrows with vegetative hedges, conservation measures with brushwood and loose boulder structures with vegetative support and treatment of drainage lines at different levels of hill slopes. Apart from this, a number of crop production measures for improving cropping system were also taken up in the watershed area.

The NWDPRA is a very big scheme involving huge amount of investment, aiming at improving the living conditions of a large number of rural people. To have a feedback regarding to what extent it has achieved its objectives and to suggest the remedial measures, if necessary, Ministry of Agriculture, Government of India asked the AER Centres to take-up evaluation studies. The job of evaluating the impact of Naravagadda Watershed was assigned to AER centre. Waltair and it was completed in the year 1996 taking 1992-93 and 1993-94 as reference years for respective programmes completed in those years.

The study of the impact of this programmes shows that there was a significant increase in the input use, yields and net profit of the farmers after the implementation of this programme. A tendency of the farmers towards change in the cropping pattern was also observed. There was an increase in the moisture content of the soil and ground water levels in the programme implemented area. A significant part of the non-arable lands was brought under horticultural crops.

Evaluation studies on Watershed Development Programme conducted at various places indicated almost the same results. The programme is very successful in achieving its objectives immediately after its implementation. The nature of this programme is such that, its impact can be seen immediately. But sustainance of such an impact in the future is not automatic since it is a low cost programme and the structures created are very small. They may disappear within no time if they are not properly maintained. After the project is implemented, it is the rasponsibility of the farmers to maintain the structures. Therefore, the impact of this project will be sustained only when the farmers realise the importance of the structures created and maintain them properly.

The studies examined the sustainability of the programme are very limited. The Agro-Economic Research

Centre, waltair conducted a Survey of Naravagadda Watershed in October, 1999 to know what is the status of the structures created under the programme and whether the benefits created by the programme are being sustained. An attempt has also been made to identify the constraints if any that stand in the way of sustainability of the programme. The data are collected from the same of households which covered in the earlier survey.

The Results of Re-Survey

The following is the analysis of the results of the re-survey.

Present Status of the Soil and Water Conservation Structures

Since there was no transparent economic benefits, farmers did not maintain the structures like bunds, vegetative hedges and drains etc. On the contrary many farmers cut the bunds to save succeptible crops from temporary water logging. Contour vegetative hedges, gully control works which come under soil conservation were partly destroyed and these were not maintained either by the beneficiaries or by the soil conservation department in the watershed area beyond the year 1996. The department of soil conservation has totally withdrew its duties soon after the completion of the duration of the project. The stone checkdams constructed in the watershed area are still in good condition and fairly useful in safeguarding the soil erosion and to preserve moisture in the fields. The soil and water conservation programmes that were implemented in the Naravagadda Watershed are acceptable, relevant and sustained. The suitability of this activity have proved by contributing the increase in ground water table. The water level in the existing dugwells increased significantly in all the watershed villages and thereby increased the production of various irrigated-dry crops.

Impact of the Scheme on Groundwater

The programme implemented about 6 years ago, give expected benefits particularly in respect of groundwater. More than half of the beneficiaries possess dugwells for irrigation. At present, the water levels in these wells increased with a range of minimum 2′ and a maximum of 8′ level. The total depth of these wells ranges between 25′ and 35′ with a diametra of 15′ to 20′ and the present water levels are maintained between 10′ and 15′. These results show that the impact of NWDPRA programme on ground water levels in the study area is sustainable.

The project was found very useful by the farmers as it helped in combating adverse affects of drought and saved their crops. With the improvement in well irrigation, new crops like sugarcane, vegetables and horticulture were introduced.

Production System

(A) Changes in Cropping Pattern

The second major component of watershed development plan is the promotion of diversified production system that can be sustained by the pattern of rainfall, availabile–moisture and farmers' resources. Major changes have taken place in crop pattern by introduction of new crops and plantation crops in larger extents during the post project period.

Horticultural crops include Mango, Cashew, Coconut, Guava etc., gain importance and the area under these crops alone constitute 80 per cent of the total cropped area of the beneficiaries in 1999 when compared to 65 per cent of the area of the corresponding crops of the same sample in 1993. The per cent or area under food grains and oilseeds has been decreased from 28 per cent in 1993 to 12 per cent in the year 1999. This change in crop pattern is mainly due to introduction of new crops like vegetables, chillies and sugarcane in kharif season and alloting more areas towards plantation crops for better utilization of treated areas and to bring marginal lands under cultivation.

The crop pattern of non-beneficiaries indicate that though horticultural crops occupy 50 per cent of the total cropped area millets and sugarcane are the other important crops in kharif. There is no shift of crop pattern towards high value crops in both the seasons.

The beneficiaries are in an advantageous position over the non-beneficiaries in having larger areas under plantation crops and cultivating other remunerative crops in the treated areas. Increase in number of wells having parennial water as well as a rise in the water level of existing wells are the main reasons for a better position of beneficiaries over non-beneficiaries.

(B) Changes in Crop Yields

Since, the production on horticultural crops is just started, the yields are not compared. The yields of annual crops mainly food crops which were grown in both the periods are compared. There is slight improvement of per hectare yields of Bajra (83 kgs/ha) and Paddy (121 kgs/ha) of the beneficiaries during post project period over project period.

Yield improvement on the beneficiary fields has been substantial as compared to non-beneficiaries for all the crops grown in 1999. (Paddy 813 kgs/ha, Bajra 615 kgs/ha, Ragi 103 kgs/ha). This wide gap in yields is attributed due to better adoption of agricultural practices on beneficiary farms. Since these areas are dominated by marginal and small farmers, they are the major beneficiaries of this programme.

Changes in Bio-Mass Production

In the context of watershed, bio-mass conservation, production and controlled use without coming conflict with ecology, serves as a potent source of fuelwood, fodder, fibre and fertilisers. The development of legumes and fodder on cultivators' fields is compeltely neglected due to erratic rainfall. Fuel-wood and horti-plantation is sustained on bunds and keylines in the watershed area and this was not practiced by the non-beneficiaries as these areas are not treated. The hill slopes and common grazing lands are the main sources of feeding the cattle. The farmers are aware of the benefits from bio-mass production and are prepared to promote the activities.

Status of Migration of People

Though the watershed area is often affected by drought, the people depend on non-agricultural employment in the nearby city of Visakhapatnam in times of hardship. The increased cultivable area, improvement in income due to increase in agricultural production and availability of employment opportunities in the villages are the indicators for non-migration. Out-migration of neither people nor cattle are reported by the sample households.

Main Findings

1. The study indicates that soil and water conservation activities are helpful in improving the irrigation potential, while contour bunding helped in reducing the velocity of run-off water and soil erosion.

2. The results of the study indicate increased dugwell irrigation leading to higher crop diversification and substitution of low value crops with more profitable crops due to watershed technology.

3. Shift in cropping pattern is noticed in favour of remunerative crops. The project was helpful in increasing wage rate and gainful employment.

4. Additional areas were brought under cultivation after the treatment of watershed and thereby

increased the crop intensity, crop production and also yields.

5. It was indicated that other programmes, such as agro-forestry, pasture development and dairying should have been given enough attention.
6. Habit of increased dependence on subsidy or outside help are said to be the reasons for non-maintenance of the watershed structures by the participants in the project area.

Policy Suggestions

1. As the NWDPRA project has a significant impact on the conservation of resources like water and soil, crop pattern, crop yields and income of the people and this impact is observed to be sustainable over time, the scheme may be continued to cover all the watersheds.
2. It is necessary to maintain the structures created under NWDPRA in the longrun for better sustenance of the development programmes. 'Watershed Committee' must be constituted in the lines of 'Water Users Committees' that exist in Andhra Pradesh and the Government has to provide matching grant and technical guidance for better management of watershed programme.

10

Agriculture and Resource Management with Special Reference to Andhra Pradesh

N. Sudhakara Rao

Introduction

Agriculture is the backbone of Indian economy. Andhra Pradesh is the 4th largest State in the country in terms of land area and population. Agriculture accounts for $1/3^{rd}$ of the GDP and provides livelihood for 65 per cent of the population. The State has the largest coast line (974 km) in the country and hence the State is subjected to severe and frequent cyclones. About 60 per cent of the land area depends on rains. Eratic rainfall, frequent cyclones, under utilization of land and other resources are the major threats to State economy.

Improvement in agriculture production, its efficiency and effectiveness are closely interlinked with environmental considerations. The subject of environmental economics is a recent field of study in the field of economics. The term environment in the present context does not merely mean surroundings. It includes all living and non-living objects, happenings and forces both natural and manmade which influence the life of an organism, in relation to the environment. The environment is a part and product of nature and is governed by natural laws and principles. Some natural occurrences can create, alter, destroy an environment. Once a form of life is last from the earth it cannot be created again.

The term environment is an interdisciplinary concept and includes a wide variety of issues concerning human settlement, natural resources, plants and animal life and pollution of all sorts. Environment denotes such natural things as the earth, atmosphere, healthy air and pure water which are essential for the existence of human life. It includes the flora and fauna and all the living things in an area and the physical elements consisting of the soil, climate, waste etc. of that area. Environmentalist recognize change as a universal, needed for the progress of human kind. Present threats to the environment are the result of the rapid technological development and injudious use of natural resources.

Resources

Environmental resources are those which are present in our environment and which we use in our day to day life. These resources are renewable and non-renewable. Renewable resources are those which are automatically renewed and which are present in limited quantities and hence if these resources are consumed injudiously, we may not find them again. Obviously the non-renewable resources are more prone to depletion. Hence, we have to be much more careful while exploiting these resources as these cannot be created if once exhausted.

Environmental Problems

While using the environmental resources, we put lot of pressure on resources whether knowingly or unknowingly. Environmental protection is considered to be one of the important things in the process of economic development.

In the modern technological world, human life faces many environmental problems. The very existence of human life is at stake unless and until solutions are found to these problems. The development affords are making way through industrialization, urbanization and excessive exploitation of resources, which have resulted in the depletion of resources. Environmental degradation should not be allowed to continue if the world is survive. Environmental problems undermine development by causing damage to productivity. Degraded soils, depleted acquifieres and eco-systems that are destroyed in the pursuit of faster rate of growth ultimately jeopardise the prospects for earning income in future.

Wet Land Management

Out of the total geographical area of 275 lakh hectares, the net cropped area is 39 per cent. The state made rapid progress in agriculture and providing food security to the people. The food production increased from 59 lakh tonnes to 137 lakh tonnes during the period 1956-57 to 1999-2000. The per capita food availability increased from 164 kg to 206 kg during the same period. The net irrigated area increased from 28.6 lakh hectares to 43.8 lakh hectares during 1956-57 and 1999-2000. Canal irrigation increased from 12.7 lakh hectares to 16.3 lakh hactares, and irrigation through well increased from 3.2 lakh hactares to 19 lakh hactares. However, irrigation through tanks declined from 11.8 lakh hactares to 6.5 lakh hactares during the same period.

One of the important components of environment both biological and physical is the wetland ecosystem. It accounts for about 6 per cent of the global land area and is considered among the most threatened of all environmental resources. In the recent past wet lands have been destroyed or altered on global basis as human society sought to exploit the benefit provided by these natural systems. The loss are occurring to wet lands due to intensive agriculture, aquaculture, industrial use hydrological perturbation, grazzing and fishing activities.

Ecologically wetlands are important ecotones which are transitional between open water and land, endowed with definite structural functional activities and performing specific ecological roles. They are water saturated and submerged areas. Wetlands have the unusual ability to upgrade waste water effluents. Most of the wetlands are more productive than tropical forests. Wet lands recycle large quantities of nutrients and sediments from the drainage basin.

Wetlands are very valuable environmental assets. Cultivable lands receiving the benefits of irrigation from river and rainfed tanks. Wetlands are an assured sources of farm production. Therefore, they have received special attention in India under the performance of achieving self-sufficiency in food production. It helped green revolution and the country achieved stability in production and self-sufficiency in foodgrains, mainly because of wetland irrigation.

Wetlands are not managed in an economically optimal way. Agricultural run-off pollution is major concern in this context. Expansions of the cities reduce the area of wetlands in its natural state. Damage is caused from pollution of river water by the discharge of sewage and other ways from the city causing severe pollution with it deleterious effects and the fertility of the soil.

There is great need to formulate policies for the management of wetlands. The hydrology of wetlands has not been sufficiently analyzed. The wetland ecosystem is not properly rehabilitated and ameliorated. Water table control, reuse and disposable of drainage water system have become another problem today.

Wetlands sufferer from problems of water logging, salinity, alkalinity of soils an account of inadequate planning and inefficient management of water from a high percentage of alkaline salts where the productivity is less than 50 per cent of normal soil. Intensive agriculture with modern chemical fertilizers no doubt is high yielding, but it also depletes the micro nutrients available. About 40 per cent of the cropped area in India is affected by floods and droughts as a recurrent annual phenomenon.

Thus, wetlands have been destroyed or altered as human society sought to exploit the benefits provided by natural ecosystems. Expansions of the cities have reduced the area and under wetlands in its natural state. Unexpected soil benefits have been sacrificed for a small tangible monetised private benefits. Wetland users are lacking guidance in appropriate economical signals which reflect the resources utilizations. There is every need to plan for wetland management in the country, so that the wetlands can be preserved and environmental issues arising out of wetlands can be controlled.

Deforestation

Forest represent a well organized and involved community of plants and animals. They provide several products of daily use such as food, timber, firewood, woodpulp apart from being a vast store of house of medicinal plants. India has a lands area of 633.4 lakh hectares notified as forest representing 19.2 per cent of the geographical area. The national forest policy of 1952 recommends that 33 per cent of the land are of the country should be under forest. Excessive exploitation of forest and over grazing have seriously decimated our forest resources. In Andhra Pradesh, the forest area increased from 58.5 lakh hactares to 62 lakh hactares, but it is still below the norms prescribed by the committee. Deforestation has severe implication on the health of the individuals.

Soil Pollution

Soil is a essential medium for the growth plants. It takes hundreds of years for the formation of top soil layer which supports plants to grow. The management and improvement of soil fertility is vital for agricultural production and economic prosperity. Deforestation over grazing, unscientific agricultural practices, desertification, led to soil erosion and soil pollution. The term pollution is derived from a Latin word pollutes which means spoil or to defile. There are many types of pollutants such as air pollution, water pollution, sound pollution, soil pollution. These pollutions come from all types of human activities viz., agricultural, industrial and household activities. Soil pollution is one of the important pollutions and considered to be a serious and dangerous burning problem which can hardly be neglected in future.

Soil pollution is usually a consequence of insanitary habits of various agricultural policies and incorrect methods of disposable of solid and liquid wastes. It also can result from fall out from atmospheric pollution. Soil pollution damages the thin layer of fertile soil which took thousands of years to be formed through the natural process.

Pesticides

Pesticides are an important input in modern agriculture. Large-scale application of pesticides such as aldrien, D.D.T. had adverse effects on health. The application of pesticides effects not only the quality of soil and water and may have undersirable effects on flora and fauna. Pesticides create environmental problems such as pollution of soil through pesticides residue, contamination of surface and ground water, extension of many insects and bird spies and enters into human food chain.

Andhra Pradesh consumes the largest quantity of pesticides in the country (75 lakh litres in 1998-99). The per capita consumption is 0.85 litres is high as compared to other countries. The indiscriminate and spurious use of pesticides add to the misery of the farmers. Recently Integrated Pest Management (IPM) and Integrated Disease Management (IDM) was introduced and these strategies are to be intensified.

Fertilizers

Agriculture production can be increased by increasing land area or by using more fertilizer. Plant nutrients are introduced into the soil either through organic-manure or through chemical fertilizer. Intensive use of fertilizers can have many undesirable side effect on the environment. Phosphates are almost observed into the soil and start to leach into the ground water after a few decades. Intensive use of nitrogenous fertilizers causes large quantities of nitrates to leach into ground water. Only 50 per cent of the fertilizer is actually used by the crop and the balance last through leach and soil run off. Thus, larger scale application of fertilizers will have adverse effects on the environment.

Andhra Pradesh is the second largest consumer of fertilizers in the country (20 lakh tonnes in 1998-99). The major nutrients NPK are applied considerably. There is excessive use of N fertilisers and farmers apply P and K which do not need them. The balanced use of fertilizer is very much essential. The Integrated Nutrient Management should be practised. Indian soils lack sufficient zinc and NPK and they should be used in proportion only.

Water Logging

Upto 80 per cent irrigation of water may never reach the crop at all. Seepage of irrigation water into the soil water will lead to rise in the water table leading to water logging. Water logging is detrimental to most crops because its reduces soil aeration. Much irrigated area require effective drainage management and measures to reduce water logging prevent water pollution.

Saline Soils

Saline soils contain a concentration of mutual soluble salts which are chlorides and sulphates of sodium, calcium and magnesium. Hence, the term saline soils refer to soils containing sufficient soluble salts to interfere with the growth of the most plants. The death of crops due to soil salinity is not only an economic disaster for the farmers, it my lead to disfertication, as soil become exposed to agents of soil erosion. Steps are to be taken to convert saline soils into ordinary soils.

Sustainable Development

The concept of sustainable development which was introduced by the world commission on environment and development (Bruntland Commission paper on common future 1987) has wider policy implications since environment and economic development are mutually dependent on one another. Development is undermined without proper environmental protection and there is every need for sustainable development in the context of the conservation and development of ecosystems. The idea of sustaining the earth has a powerful argument for raising public awareness and focusing on the need for the better environmental quality.

As per the Bruntland Commission sustainable development means "meeting the needs of present generation without compromising the need of the future generations". It is about sharing of well being between the present and the future generations. Sustainable development is one that increases the welfare of the current generation provided welfare in the future does not decrease.

It is not plausible to preserve all natural resources. Successful development will inevitable involve some amount of land clearing, oil drilling river damming and swamp draining. The net effect is no loss or depletion of natural capital. In the past, the benefits from human activities have been exaggerated and cost of environment have been ignored. Now, there is every need to balance between development for the present and the future as far as possible.

There is every need to have sustainable development. One of the solutions of the environmental problems lies in controlling population growth. The affluent trap face complex environmental problems as a consequence of applying science and technology without much regard for environmental consequence. Many people in development countries live below poverty line. The situation of poverty trap becomes much more severe in the years to come unless and otherwise effective steps are taken. In Andhra Pradesh, the people below the poverty line decreased from 48.9 in 1973-74, 16 per cent at present. Still there is every need to minimum absolute number of people living below poverty as it did not decline.

Summary and Conclusion

In Andhra Pradesh, the agriculture has developed well during the last five decades of development. However, the percentage of people depending on agriculture has declined from 69 per cent to 65 per cent during 1960-61 and 1990-91. Agricultural development is closely associated with environmental pollution. The use of land for irrigation, use of pesticides, fertilizers and methods of intensive use of cultivation has lead to environmental pollution of different types. Land and water resource management are very essential in the present context. We should not use underground water indiscriminately. As per experts opinion, we should not use the ground water more than 10 cubic metre depth. There is every need for watershed management. There is every need to put the agriculture on sustainable basis. We should not use the land and water resources indescriminately. We should always keep the future generation in view while exploiting land and water. There is need for diversification of agriculture so that land fertility can be protected, otherwise the land losses its value in near future. Crop rotation will help in the better management of both land and water resources. There is every need to adopt strategies for better management of agriculture in the country and in the State.

11

Agriculture and Environment: Emerging New Issues

B. Sambasiva Rao,
Y. Indira Kumari
M. Babu Rao

The rather facile optimism of the 1970's about the prospects for agricultural growth based on Irrigation-HYV fertilizer technology has become muted in the face of experience. Increasingly and this reflects a worldwide surge of concern and interest in environmental issues–the deleterious consequences of a single–minded pursuit of this package are becoming more prominent in public discussion of agricultural strategy. The concern centres on the need to arrest and reverse the process of land degradation, the wisdom of investing in large surface irrigation projects, over-exploitation of ground water, the reduced bio-diversity and the excessive use of fertilizer and pesticides.

Land Degradation

The progressive depletion of forest cover is assuming alarming proportions. Recent assessments based on satellite imagery suggest that the extent of open forests–which are too thinly stocked with trees–is much smaller than recorded in official forest statistics. Deforestation of the hills and upper catchments of major river basins increases erosion, intensifies surface run off and reduces the duration of stream flows. Over grazing and cutting of trees to meet fuel wood needs of a rapidly growing population is causing a similar problem in rural settlements. In land degradation, as estimated 166 million hectares (out of total surface area of 235 million hectares) in India are assessed to be "degraded", about 105 million hectares on account of erosion, some 32 million hectares due to depletion of nutrients and salinity, 7 million hectares are due to waterlogging and 38 million hectares are stable terrain and some 22 million hectares are unfit for any agriculture. There have been two major initiatives in the past two decades to tackle this problem. One is watershed development and the other is the creation of the National Wasteland Development Board. The watershed programme in turn has two distinct components. One deals with afforestation and other soil control measures in the catchments of major reservoirs and the other is the acceptance of integrated development of land, water and vegetation in micro watersheds. It has already noted that the acceptance of the concept of integrated watershed planing has not been translated in practice on the ground.

Status of Large Irrigation Projects

Criticisms levelled against larger irrigation projects reflect several different, but related issues including the high cost, low utilization, submergence of variable land and displacement of people and degradation of land due to indiscriminate irrigation. Large reservoir based works are expensive and costs have risen steeply. Large cost and time overruns are common. In this juncture, an enquiry by the public accounts committee of parliament in 1983 reported that the final cost of some forty projects, which were reviewed, turned out to be two and half times the estimated cost, the difference ranging from 65 per cent to 67 per cent.

The average time taken to complete projects was more than twice the target, 14 years against 6 years. Costs of large projects are indeed high and rising. This is due to the laxity of their planning, scrutiny and implementation. Poor preparatory investigations, delays in completion due to inadequate funding and lack of any pressure for economic use of time and resources have become chronic features of public sector irrigation programmes. In relation to the problem of canal irrigation areas, water logging and salinity arises from indiscriminate irrigation, poor drainage and rise in water table due to accumulation and see page underground. Estimates of area suffering from these problems range from 12 million hectares to 20 million hectares, but there are no precise, systematic measurements of their incidence nor of their direction and pace of change overtime.

Groundwater Regulation

The lowering of the groundwater table is much documented. In most of the states, water level has fallen by more than two meters over a short period and in many areas the fall is over four meters. This is accompanied by an increase in flouride levels in groundwater. But, there are no reliable estimates of the recharge rates and the actual rates of extraction of groundwater in different parts of the country for areas commanded by surface systems and for rainfed areas. Therefore, it is difficult to judge whether and where the current rates of exploitation exceeds the sustainable levels. The lowering of the water table following the increase in the number of wells and their energisation is clearly a signal to be heeded, particularly when it is accompanied by a lowering in the yield per well or increased incidence of wells going dry altogether.

Competitive deepening of wells increases both private and social costs of water and yet there is great unwillingness on the part of farmers to try out cooperative ownership and operation of wells. People seem to have strong preference to own their wells if they can afford it. There is clearly a social myopia on the part of the users in failing to see the longterm consequences of progressive fall in groundwater table and the need for collective regulation to avoid a predictable and eventually disastrous fall in output and income. State regulation is the solution uppermost in the agenda. In principle, regulation by the local community which has a direct and immediate stake in the proper use of the resource would be more effective, besides being more economical and equitable.

Chemical Pollution

If over exploitation of groundwater and indiscriminate use of canal water leads to the salinity problem, the growing use of fertilizers leads to chemical pollution of water resources. The use of nitrogenous fertilizers produce nitrates which end up in streams and groundwater reservoirs. Continuous high intensity fertilizer-use leads to a progressive increase in the nitrate content of confined water bodies (particularly groundwater). Though, the overall intensity of fertilizer use in India is low on the average, there are parts of the country where the intensity is high. By 1985-86, 48 districts (concentrated in Punjab, U.P., A.P., Tamilnadu and Karnataka) reported an average application of 100 kg nutrients per hactare. There are reports from the Punjab of the nitrate content of groundwater increasing to unsafe levels. Chemical pollution due to intensive use of pesticides is also a growing hazard. Again though the overall level of use is low (377 grams per hactare compared to above 1500 grams in developed countries), the intensity in some pockets is high. Another concern is the extensive use of pesticides which have been banned as hazardous in other countries.

The problem will become more widespread and serious as fertilizer use intensity continues to rise, systematic monitoring is essential to get a better idea of magnitude and location of the problem and how fast it is spreading. At the same time this predictable danger reinforces the case for fertilizer conservation through more careful use of chemical fertilizers and through measures to augment the use of organic substitutes such as farmyard manure, human waste and natural nitrogen fixing agents.

The introduction of new varieties (especially HYV) has sharply reduced the number of locally adopted varieties the high yield potential of HYVs is thus bought at the cost of much greater potential vulnerability to disease and pest attacks. Breeding pest and disease resistance is now part of the varietal improvement programme and agronomists are trying to propagate integrated pest management through a combination of resistant varieties, chemical and natural biological control mechanisms present in local environments.

The wisdom of breeding for high yield at the risk of reducing biodiversity, the efficacy of gene banks as a means to preserve genetic diversity which can be tapped wherever needed, the location, control and management of these banks to safeguard free access, the question of the property and patent rights over new strains bred from varieties evolved through a long process of selection and trial by traditional societies, are some of the issues which have come to the fore in the last decade.

12

Corporatisation of Agriculture in Andhra Pradesh: A Review

N. Vijaya Ratnam

Agriculture sector in the State of Andhra Pradesh has multifaceted role to play and it should not be misconstrued by the fact that it is the just source of livelihood for the great majority of population. But agriculture's pervading multitudeness should be understood in a diversified manner from manifold facts: (i) agriculture forms the basis of state's industry including trade and transport, agriculture holdings, farm inplants, irrigation works and livestock represent the largest fixed capital in the state, (ii) as an important source of income, according to the Estimates of States income for 1996-97 at current prices, agriculture alone–contributed 27.8 per cent of the total income of the state, (iii) about 73 per cent of the population of the state lives in rural areas, and (iv) as revealed by 1991 census about 65 per cent of the population derive their means of sustenance wholly on land.[1] Besides, there was a record agricultural production of 149.05 lakh tonnes in the year 1998-99 even as 12.76 lakh more hectares were brought under cultivation.[2] The objective of this paper is to review the developments in Agriculture in view of the

Government's current policy on Agriculture more specifically on corporatisation of Agriculture in Andhra Pradesh.

Corporatisation of Agriculture

A new tendency of farming has heralded in Indian Agriculture in the recent past in a big way is contract farming, company farming or corporatisation of agriculture with the objective of diversifying production, export markets and benefitting the farmer with comparatively higher rate of returns. Under this system, private agro-processing companies provide extension services, related inputs and management practices after entering into buy-back agreements with fruit and vegetable growers, cultivators of cash crops such as tobacco, sunflower, meat and dairy producers.[3] The mechanism and economics is built that there will not be any loss either to land or right of ownership of the farmers.[4] It is not out of context to mention that the Tata Economic Consultancy Services[5] (TECS) has called for introducing corporate farming in horticulture, oilseeds and foodgrains. This institution sees private and public investment as complimentary, with former driving food processing and biotech and the latter seeding growth in untapped regions while continuing research and developing thrust.

To quote a few practical examples, this system has been pioneered by none other than Pepsi Food Company in Punjab by entering into contract with farmers initially growing tomatoes and followed later by chillies and potatoes. As part of the implementation package, company has provided good seeds and other technological inputs along with new technology to the farmers. The outcome of this experiment has been a rousing success as this arrangement has benefitted both farmers and the company.[6] Besides, the Pepsi Food Company has fertile market for its agro–processed products in different world countries including USA, China, Middle East, Europe, Canada, Nigeria and

Bangladesh with an exports standing of Rs. 375 crores in 1998-99. Having experienced a rousing success, the Pepsi Food Company is raring to go with a plan to replicate the system in Gujarat and Karnataka.[7] However, contrasting experiences in this type of farming is not difficult to visualise and as it is not a remote possibility.

Andhra Pradesh Scenario

As part of the new Agriculture policy to develop and bring about radical changes in the agriculture sector of Andhra Pradesh, the Government has envisaged that contract or company type farming should be introduced. Speaking at the valedictory function of workshop on 'Carving an Agenda' for the Federation of Farmers Association, the Commissioner of Agriculture, Mr. Ajay Kallam, said that already production in some sectors like sugarcane, oil palm and seeds were entirely on contract farming way[8] and it is no way a new scheme of things in Andhra Pradesh's agrarian culture.

The Government of Andhra Pradesh as a part of its programme of agricultural development to utilise the hitherto neglected and disadvantaged natural resources, for the benefit of the small peasant community who possess uneconomical and small landholding has proposed a scheme in the direction of the contract farming or company farming. The Vision 2020-Andhra Pradesh's mission envisages that to enable its farmers, entrepreneurs and professionals to make agriculture flourish, corporate type of farming or corporatisation of agriculture is a necesity. The Large Scale Advanced Farm Project (LSAFP) which is better known as the Kuppam Project[9] has a wholistic approach to modernise agriculture by involving agriculture and all other departments in the process of technological upgradation. This project revolves around the Isreali cooperative model of agriculture. This project was started in June, 1997 on a cooperative basis by involving 167 small and marginal farmers with an acreage network of 170. The varied types of inputs required for start up of the project have been provided by the Government Departments and Company sources such as Japan International Company agency and B.H.C. India Limited affiliated to entities located in Israel and U.S.A. which have a larger stake in participation and income sharing. It is rather an expensive project which incurs an estimated amount of Rs. 9.63 crores, out of which the State Government has to incur an amount of Rs. 243.5 lakhs exclusively for the installation of the know-how besides contributions from small and marginal farmers to the extent of Rs. 1.52 crores.

It was expected that Kuppam model would provide the demonstration effect which will enable replication of this model in other places too. The purpose of the large scale project was to raise irrigation efficiency, increase productivity and implement demonstrate the state of art agricultural technology and methods being used in Isreal. The improved technologies include.

- detailed analysis of soil and water;
- design and installation of advanced irrigation system to utilise every drop of water;
- mechanisation of agriculture; and
- new technologies and modern practices of growing.[10]

Expected Outcomes

The following is the list of anticipated and expected outcomes of the project from the viewpoint of farming community:

- Providing agricultural infrastructure to the less developed lands;
- Upgradation of agricultural practices and systems;
- Creation of base to farmers and systems;

- Increase farmers outlook and approach;
- Improve skills and living standards; and
- Ensure socio-economic development.[11]

However, the company too has its own expected outcomes in the form of infrastructure establishment such as food processing units, cold storage units including transport, creation of market place for exportsand industrialisation of food processing sector.[12]

A study group[13] has made certain observations to show whether the contract farming has yielded the expected outcomes from the point of view of farming community. Though Kuppam Project has brought highest technological breakthrough in the era of Green Revolution under company type of farming by inducting columns of machinery to dig or drill the feet of small farming community for some partial advantage at the moment, the future positive consequences of this mamooth exercise are very much bleak. A host of undesirable consequences persist at the moment though at the gains of returns at the expense of heavy investment by the state and farmers combine can be enumerated as follows. The observations made by a group of social scientists from a reputed research institute are referred to herein.

- It is observed that the project is run on a secret deal between the State Government Departments and Isreali technical group without dissemination of the details of the farm management to the members itself who are small and marginal farmers.
- It is observed that this type of farming has no chance of participation of the member farming community except hiring them as agricultural labourers or supervisors as and when situation demanded by the project management.

- The project has cultivated more of commercial crops which have international market potential as compared to a few food crops which have been observed to be spotted in a limited area.
- An observation of the company type of farming reveals that many unscientific methods of cultivation have taken place. To quote a few:

  (a) Unscientific and disproportionate use of chemical fertilisers and pesticides has resulted in not only severe damage to the land but also environmental degradation. The evidence indicates that out of the total cost of production of Rs. 20,000 per acre for each year about 41 per cent has been invested towards chemical fertilisers and pesticides alone.

  (b) A tendency has been set in the company type of farming that only a few members of the family are getting gainful employment leaving many jobless and even without a substitute occupation. The most saddening fact is that the system has culminated into a situation, wherein the farming community has to purchase even essential commodities like rice, cereals, milk and firewood from the market leaving the people very much in the danger of food security.

  (c) The groundwater exploitation under the company farming system has heavily damaged availability of groundwater to the surrounding areas. It is evident from the fact that company has digged borewells to the extent of 600 feet depth as against a normal depth of 250 feet digged by the surrounding farming community.

  (d) The possibilities are immense that farmer will be separated from land and land rights as

some of the members have started selling away their landholdings to the company at a lower price. In the worst situations of company collapse and cease of contract and the farming community will be subjected to lot of hardship in terms of payments as it is very much evident from the recent happenings in the case of non-payment of dues to sugar and palm oil farmers.

(e) Another important observation has been that there is widening gap between labourers under company farming system and non-company farming system in terms of wages and hours of work. It is dissatisfying to note from the study that workers under company farming system are paid wages 12 per cent less than wages paid to workers under normal conditions. Further, it is also dissatisfying to note that labourers under the new system are subjected to lot of exploitation as they have to work for a lenghty period 11½ hours, however, wages were paid only for 8½ hours. Besides gender discrimination in respect of payment of wages has always been a permanent feature.

(f) Contrary to the expectations, the high rate of return from the company farming is not greater than the non-company farming. It is evident that under corporate farming of Kuppam each acre yields an income of Rs. 13,100 at an investment cost of Rs. 40,000 per year. However, under non-company farming a farmer in irrigated conditions or rainfed conditions could easily earn an equivalent income with much lower and normal proportion of expenditure/investment,

by raising a food crop and a commercial crop subject to minimum support price being provided by the Government. Further, it is noticed that the amount earned by the farming community is just sufficient to service the debt raised to contribute towards company type of farming.

The listed findings on Kuppam Cultivation Project drive home a pertilent point that the outcomes expected from the project did not materialise or have not been achieved. The wishful thinking of the State Government that such type of projects should be emulated in different parts of the state with all the natural disadvantages has not gained any ground as no evidence in this direction available so far, though some of the projects are working not at the cost of the exchequer to the Government and not without keeping the farming community alienated.

It is not out of context to put forth the viewpoint that the policy of the Government to privatise the state level public sector undertakings and promote corporatisation of agriculture by inveing huge amounts in the corporate type of farming casts a shadow on the Government that it is maintaining double standards as between agriculture sector and industrial sector.

Conclusion

In order to curtail the ill-affects associated with corporatisation of agriculture as specified in the preceding paras, the following suggestive measures have to be considered:

1. Redical redistribution of land among the landless, forms an essential precondition for the sustainable and healthy growth of agriculture in any agro-based country like India. It is only by this perception, the policy-framers, since the dawn of

planning era in the Independent India, have offered a big seat for the event of distribution of surplus land for the rural poor. So far as Andhra Pradesh is concerned, only a paltry figure of to lakh acres have been distributed during the past 50 years. Hence, no positively significant growth has taken place, as observed by the Noble Laureate Prof. Amartya Sen.

2. Small farms have been proved more productive than the big ones. Hence, continuation of the cultivation under small farms in the Indian context also forms an essential prerequisite. As Dr. M.S. Swaminathan, India's Visionary Scientist has very rightly observed that small farms are boon but our small farmers are weak and hence need to be supported by the State.

3. The projects of corporate nature which involve small and marginal farmers in agriculture sector in the state in the state of Andhra Pradesh have to be established on a long-term basis by allowing adequate participation and involvement so as to get imparted with the innovative methods and skills to manage their own farming affairs without the aid of foreign technical people.

4. Rejuvenation of Farmers Training Centres is the need of the hour. Sequel to this step, measures have to be initiated by the Ministry of Agriculture to chalk out specialised Training programmes by appointing the specialised faculty with experience in adopting advanced and upgraded agricultural cultivation practices from within the country.

5. It is not out of place to put forth that the Government of Andhra Pradesh has to provide finance in liberal terms, to those farmers who augment cultivation in the naturally

disadvantaged areas/regions so as to enable them to switch on to cultivate new commercial crops with highly demanded market by adopting modern technical know-how.

REFERENCES

1. Government of India, "Andhra Pradesh 1999-2000, Small Industries Service Institute, Ministry of Small Scale and Agro and Rural Industry, Hyderabad, p. 10.
2. *Ibid.*, p. 10.
3. Rita Sharma, "Role of Extension in Indian Agriculture", *Financing Agriculture, Special Millennium Issue,* October-December 1999, p. 71.
4. The Hindu. February 19, 2001, p. 5.
5. Economic Times, February 23, 2001, p. 3.
6. Dattatreyulu, M. "Agricultural exports from India: Scenario and Directions", *Financing Agriculture,* October-December 2000, p. 22.
7. *Ibid.*, p. 23.
8. The Hindu, *op.cit.*, p. 5.
9. The Hindu, December 31, 2000.
10. *Ibid.*, p. 8.
11. *Ibid.*, p. 8.
12. *Ibid.*
13. The observations on company type of farming made by group of scientists of a research institute are published in a book entitled "Agriculture System in Andhra Pradesh—For Whose Sake" by Sundaraiah Viganana Kendram, Hyderabad.

13

Productivity Management in Agriculture—Modern Approaches

A.K.V.S.S.R. Raja Sekhar

Agriculture accounts for one-fourth of India's national income. About 64% of India's workfore is directly engaged in agriculture for its livelihood. It still accounts for sizable export earnings. Nearly 80 per cent of India's poverty population lives in rural areas and is directly or indirectly dependent on agriculture. India's water resources, both surface and ground water, when fully exploited would bring half the cultivated area under irrigation. The abundant sunshine offers considerable scope for increasing crop yield as well as for harnessing forest resources as renewable sources of energy. However the application of modern technology to agriculture is still lagging behind the advanced agricultural systems of the West as well as of the East. The successful application of such technology would, therefore, increase crop yield substantially. Thus, agriculture in India offers considerable potential, both for achieving self-reliance in food on a sustained basis and for eradicating poverty.

Agricultural Growth—Institutional Changes

The growth of the food grains output was negative during the fifteen years prior to independence. The colonial system perpetuated traditional technology in agriculture except in the case of little export-oriented commercial crops. The development of infrastructure through irrigation, roads, markets etc. was extremely limited. The intermediaries in agricultures in agriculture, like zamindars and jagirdars, had grown in the colonial period in large parts of the country, denuding the investible resources and sapping and motivation of actual tillers of the soil. Thus at the time of independence, we inherited a stagnant agriculture with a deficient infrastructure, traditional technology and outmoded exploitative institutions. Thanks to the improvement in public health and the rapid reduction in infant mortality, we were faced with the problem of population growth outpacing the availability of food. The five year plans were, therefore aimed at improving the infrastructure through irrigation, stepping up the use of fertilizers and improved varieties of seeds and effecting institutional changes such as the abolition of intermediaries, imposition of ceilings on landholdings and the distribution of surplus land among the landless, supply of credit through the co-operatives ensuring remunerative prices for agricultural commodities etc.

Agricultural Growth—Role of Technology

It was becoming clear by the mid-sixties that there was no alternative to technological change in agriculture for achieving self-sufficiency in food grains. After the mid-sixties, apart from the introduction of High Yielding Varieties (HYVs) of seeds for wheat and rice, public investment in agriculture was stepped up significantly. The technology raised the profitability of investment for the farmers. Besides, in agriculture, there is a high complementarily between public and private investment. As a result, the annual gross capital formation in agriculture, public and private combined, at constant prices; rose much faster than in the pre-green revolution period.

It is also heartening to find that many of the states, where poverty is widespread and where the growth of the food grains output had slowed down in the decade of green revolution ex. Assam, Bihar, Orissa, Madhya Pradesh and West Bengal, have shown a much better performance in the last decade. Their performance is nearer, or even higher than the all India average. This encouraging performance can be attributed to the spread of new technology.

The gains in foodgrains output in the last decade have come essentially from the improved utilization of the available infrastructure and from the increase in yield per hectare. Already during the first decade of the green revolution, increase in yield per hectare accounted for as much as 80 per cent of the increase in food grains output. During the last decade, there was some decline in the area under food grains. The entire output growth in this period can therefore be attributed to increase in yield per hectare. This under scores the growing importance of yield increasing technology. The new technology increased the productivity of the inputs, infact, one-fourth of the observed in the increase in the output of foodgrains can be attributed to the increase in the productivity of inputs.

Table 1 shows that over the period 1950-51 to 1998-99, yield per hectare of all foodgrains has increased by three times from 555 kgs per hectare in 1950-51 to 1,620 kgs per hectare in 1998-99. Most significant increase has been recorded by wheat with its yield increasing from 655 kgs, per hectare in 1950-51 to as high as 2,583 kgs. per hectare in 1998-99. Productivity of coarse cereals–jowar, bajra and maize has risen relatively slowly. Most disappointing has been the performance of pulses. In fact, productivity of pulses was only significantly higher in 1998-99 as compared with 1960-61.

However a comparison productivity levels in Indian agricultural with the levels in other countries shows how low the productivity in Indian agricultural is. Such an exercises has been attempted in Table 2.

Table 1: Yield Per Hectare of Major Crops

Crop	*1950-51*	*1960-61*	*1971-72*	*1985-86*	*1997-98*	*1998-99*
1	**2**	**3**	**4**	**5**	**6**	**7**
Rice	668	1,013	1,141	1,552	1,900	1,928
Wheat	655	851	1,380	2,046	2,485	2,583
Jowar	353	533	460	633	697	873
Bajra	288	286	452	344	791	758
Maize	547	926	900	1,146	1,711	1,755
Pulses	441	539	501	547	567	622
Total Food Grains	**552**	**710**	**858**	**1,175**	**1,552**	**1,620**
Oilseeds	481	507	546	570	816	944
Cotton	88	125	151	197	208	223
Jute	1,044	1,183	1,255	1,710	1,978	1,861

Source: Indian Economy, S.K. Mishra and V.K. Puri, Himalaya Publishing House, Bombay, p. 408..

Table 2: Productivity of Land in some Countries, 1998

(100 kgs per hectare)

Crop	*Productivity*	*Crop*	*Productivity*
Wheat		Cotton (Lint)	
France	76.0	China	8.4
China	6.7	U.S.A.	6.9
U.S.A	29.1	Pakistan	5.3
India	25.8	India	3.0
Rice (Paddy)		Groundnut (in shell)	
Japan	62.2	U.S.A.	29.9
China	60.6	China	25.8
Indonesia	41.7	Argentina	24.1
India	28.9	India	10.3

Source: Indian Economy, S.K. Mishra and V.K. Puri, Himalaya Publishing House, Bombay, p. 409.

In Andhra Pradesh also the institutional measures such as abolition of zamindari system and tenancy reforms were introduced. Though the land ceiling act was also made to distribution of surplus land has not been satisfactory. Besides the institutional measures the new agricultural strategy has been vigorously implemented in A.P. Resulting in the increase in productivity of different crops over period. Table 3 shows the productivity levels of different crops in A.P. in comparison to the Indian average and also the world. It is obvious that the A.P. average is more than the Indian average in respect of paddy, maize, sunflower, sugarcane, onions, and grape. However the fact remains that the Indian average it self is for below the world levels.

Table 3: Comparison Average Yields of Different Crops in A.P. State, India and World

Crop	*Unit*	*World*	*India Average*	*A.P. Average*
Paddy	Mt/ha.	10.269	1.928	2.812
Groundnut	Mt/ha.	2.828	1.210	1.082
Maize	Mt/ha.	9.478	1.755	3.328
Sunflower	Mt/ha.	2.888	0.584	0.634
Cotton	Mt/ha.	1.065	0.223	0.202
Sugarcane	Kgs/ha.	136,515	72,560	77,231
Tomato	Mt/ha.	467	15	8.0
Onion	Mt/ha.	82	11	15.50
Grape	Mt/ha.	20	18	37.3
Milk	Kgs/animal	6815	877	700

Source: Strategy Paper of Agriculture and Allied Departments, January 2001.

Table 4 shows that there is still a wide gap between the potential and actual productivity in India. By adopting appropriate strategies, it should be possible to push up the levels of productivity to the maximum, this involves sustainability and other modern techniques of management of various resources.

Table 4: Potential and Actual Productivity

Crop	*Potential*	*Actual (1998-99)*
Rice	4000/5810	1928
Wheat	6000/6800	2583
Jowar	3000/4200	873
Maize	6000/8000	1755
Groundnut	2000/3000	1210
Cotton	700/850	223
Jute	2500/3000	1861
Sugarcane	96000/112000	73000

Source: Indian Economy, S.K. Mishra and V.K. Puri, Himalaya Publishing House, Bombay, p. 413.

Sustainability

The growth performance however, is by no means smooth. There has been environmental degradation and pollution. Unlike in the intensive systems of agriculture where pollution of environment has been caused by the intensive use of chemical inputs, the ecological degradation in rural India can be attributed mainly to the extension of cultivation to the marginal and sub-marginal dryland and to deforestation. In India, the proportion of rural population below the povertyline is highest for areas not irrigated and declines with the increase in irrigation. Therefore, the degradation of the already fragile environment in dry areas has further accentuated the poverty problem.

The causes of deforestation and degradation of drylands are many. Apart from population growth and poverty, the pressure of demand for fuel wood, timber, meat and wool from high-income segments, including urban areas, has been a major cause of degradation of the environment.

However, the rural poor have been the prime victims of this ecological degradation. This is because they derive a significant part of their income from village commons and forests in dry regions. Even though the proportion of income derived from such sources is greater for the rural poor than for the non-poor, the latter are responsible for the greater share of denudation of forests. As a result of environmental degradation, the rural poor, particularly women and children have to walk much longer distances to collect fuel wood and fodder and drinking water has become more scarce owing to the decline in the water table.

There is a clear interrelationship between the levels of agricultural development and rural poverty, on the one hand, and ecological degradation on the other. Across different states in India, the extension of areas under cultivation and denudation of forests seem to be high where the progress of yield-increasing technology is slow. In such regions, the levels of agricultural income and wages are low and poverty levels are high. Similarly, the pressure from animals such as goats and sheep on forests and common lands has been increasing in regions where growth in crop production is slow. This is because the rural poor supplement their incomes by rearing these animals.

This experience suggests that apart from yield increasing technology for crops the use of new biotechnology in forestry and Animal Husbandry and improvement in management practices like Pest Management, Weed Mamagement and Land-Water Management, is not only significant in increasing productivity but also in ensuring sustainability.

Integrated Pest Management (IPM)

IPM is nothing but controlling the pest complex through cultural, mechanical, biological and chemical means. IPM is a broad-based ecological approach in crop production duly controlling the organisms causing injury to crop production by conserving the beneficial insects.

The introduction of high yielding varieties in large areas and intensive cultivation carried throughout the year for getting maximum yields resulted in the increase of pest complex. The cultivation of HYV with high fertilization contributed to excessive pest and disease build-up. To check effectively this pest problem, many chemicals were recommended and used by the farmers as prophylactic and curative measures. Due to indiscriminate and intensive use of pesticides, the following problems arise:

(i) Development of maximum resurgence in pests for most of the chemicals.

(ii) Minor pests emerging as major pests and effecting the crops in larger areas.

(iii) Disturbance in natural eco-system.

(iv) Destruction of beneficial parasites and predators.

(v) Secondary pest outbreak.

(vi) Residual toxicity in food crops posing serious health problem to human and live stock population.

(vii) Reduction in profitability due to high cost of chemicals equipment and labour etc. have cropped up.

To overcome all these problems, a shift in policy has lead to application of IPM system, which has gained greater importance in several countries. IPM system is pest management system by which need based pest management is advocated. IPM involves utilizing all possible and available methods to keep the pest population at levels below those causing economic injury. IPM helps in reducing the use of poisonous chemicals against the pests thereby avoiding pollution in the environment and reducing residual toxicity in the agricultural produce. The IPM methods are cheaper, scientific and low cost technology methods. These methods are applicable at various stages of crop growth from sowing to harvest. By utilizing the methods of IPM, the intensity of the population is kept below the economic threshold level so that, there is no economic loss to the farmer due to pest incidence. IPM involves utilization of all suitable Techniques i.e. cultural, mechanical, biological and chemical control measures in a compatible manner to reduced pest population and to maintain the pest population levels below those causing economic injury.

Weed Management in Different Crops

The strategy of effective and longterm weed management in field crops should be the judicious integration and manipulation of whole series of agronomic practices from field preparation to harvest, so as to favour the crop and to suppress the weed growth i.e. destroying their underground propagules, by preventing their emergence and by summoninizing the effect of the competition/ interference.

Critical Stages of Weed Competition

Removal of weeds at a particular stage and keeping the field weed free up to certain early period of crop age is more important than complete weed free situation or repeated weeding up to harvest. Removing of weeds earlier or later parts of the critical period may not be so effective. Such critical stages or weed free periods from sowing have been identified for several crops under or Agro-climatic conditions.

Modern Approach

The modern approach to weed management comprises of the following measures:

1. Use of all available methods in combination to control weed successfully.
2. Damage to the environments should be as little as possible in this process.

3. Identification of the weeds before taking any control measure.

Effective control of weeds by using herbicides depends largely on the soil types, soil moisture, humidity and atmospheric temperature at the time of application.

Water Management—Watersheds

The watershed approach is devised fundamentally to deal with water resources, its quantification and distribution in time and space, so as to formulate strategic schemes, to use water most logically matching its availability against the needs. It enables framing of a realistic picture of water availability and its limits on a reasonably defined geographical area so that people may themselves decide as to how to limit their own material aspirations in relation to distribution and use of water resources. Water in its natural form being the basic constituent for agriculture, livestock, forestry and various other activities that people choose to purse life patterns and survival, its imbalance create corresponding imbalances. Thus far, water has been regarded as ample and copious to support all these activities. Its shortage felt only in recent times compels one to take a cautious, calculated and concerned approach to use it judiciously and in a sustainable manner.

The watershed it is possible to monitor water resources. Beneficial use of water in the year found economic activities, catering to the domestic water need of the inhabitants, controls for its secured use and drought mitigation etc. is possible as the system is smaller. Watershed management is holistic development concept the encompasses the management of land, water, agriculture, forestry and all related matters in relation with the people who inhabit the area and their socio-cultural system.

The rainfall is not uniform throughout the year. Monsoon rain is utilized immediately for agriculture in Khariff (July-October) cropping season. Excess moisture is retained in soil to support Rabi (November-March) cultivation. Some water is stored in the surface water reservoirs to provide irrigation during critical needs. Surface and groundwater of the watershed is also protected for domestic use and for livestock during summer. For this purpose water storage and conservation facilities are created at strategic location and utilization of water for various purposes is regulated through proper motivation measures at the social levels. Water demands for various needs too are similarly regulated since rainfall status determines the choice of crops, irrigation trends, domestic water use habits et. Occupants of the watershed learn about these limitations through experience and collective wisdom.

In A.P., several measures have been taken to improve the productivity and production. The recent management techniques of pest management, weed management and watershed are also being followed on a large scale.

14

Functioning of Rythu Bazars—A Qualitative Analysis

M. Raja Sekhara Babu
B. Sambasiva Rao

Indian agriculture has witnessed a profound change since independence with increased production of various agricultural crops such as cereals, pulses, fruit and vegetables, agricultural marketing and processing have become a more complex problem to the farmer. He has to dispose of the marketable surplus on the one hand and ensure a remunerative price for his produce on the other.[1] Mere increased production has no meaning unless the farmer gets remunerative price and prompt and immediate payment is ensured, as the motivation to produce more and more comes from the prices and ready cash that he is able to get for his produce.

The development of an efficient system of agricultural marketing can act as a catalytic agent for increasing agricultural productivity and farm income. The promotion of efficient and development oriented agricultural marketing systems in the developing countries depend, to a large extent, on the government policies.[2] To improve the situation of the farmers in Andhra Pradesh, the Government of Andhra

Pradesh have introduced the system of "Rhythu Bazar" from January 26th 1999;[3] where the farmer can sale his products directly to the consumers in the market without having any mediators in between the seller and farmer. It is progressive in keeping in view to assist for remunerative prices to farmers of vegetable producers and enabling regular supply to consumers. The concept is novel in the sense it avoids middlemen, thereby both the consumers and farmers are benefited in their own way. With the introduction of 'Rhythu Bazars' the prices in general vegetable markets come down to a greater extent. The state is providing infrastructure to the 'Rythu Bazars' arranging Market facilities and proper transport facilities at a cheaper rate and at convenient times. To say in few words Rythu Bazars is a boon both to farmers and consumers.

Table 1: Statement Showing the Functioning of Rythu Bazars

Item	*Good*	*Moderate*	*Worst*	*Total*
1. Quality	25 (33.33)	40 (53.33)	10 (13.34)	75 (100.00)
2. Prices	45 (60.00)	25 (33.33)	5 (6.67)	75 (100.00)
3. Availability of all varieties of vegetables	20 (26.67)	30 (40.0)	25 (33.33)	75 (100.00)
4. Weights	30 (40.00)	35 (46.66)	10 (13.34)	75 (100.00)

The objective of the paper is to study about to functioning of Rythu Bazars through which the consumer is benefited or not.

The town of Mangalagiri in Guntur District of Andhra Pradesh was taken for the study. About 75 consumers were selected for the study who regularly purchase the vegetables from the market. From the Table 1 we can see that 33 per cent of the beneficiaries are of the opinion that the vegetables supplied in the Rythu Bazars are of good quality, and 53 per cent have said that they are moderate,

With regard to prices 60 per cent or 45 consumers have expressed that the prices are reasonable when compared to the regular markets, and 33 per cent have said they are moderate. But 40 per cent of the consumers said the availability of all varieties are not so good. Lastly major portion of the people have accepted that the weights are moderate.

Conclusion and Suggestions

On the whole, the introduction of Rythu Bazars in the Agricultural Marketing is a newer concept and benefitted the consumers to a greater extent. The consumer are getting the vegetables of better quality at reasonable prices. Not only that the farmers too are benefitted to a greater extent because there is no middlemen in the market as the farmers are directly selling their products. But the Government have to supply all types of vegetables with good quality in the market. However, the introduction of Rythu Bazars is a boon to the people of Andhra Pradesh. The following suggestions are forwarded for implementation:

1. Supply of all types of vegetables in the market.
2. Grading of goods should be maintained.
3. The vegetables should be supplied even in the evening times.

REFERENCES

1. Subramanyam, R. (1994) : COSAMB; the national link in farm marketing, Kurukshetra, November, p. 25
2. Dr. Lahiry, S.C. (1994) : Marketing and Price Policies of Sugarcane, Kurukshetra, November, p. 17.
3. Shobhanadreeswara Rao, V. (1999) : "A beginning of new era in the Agricultural Marketing" in Eenadu Daily, Friday, January, 29.

4. Kahlon, A.S. and George, M.V. (1985) : Agricultural Marketing and Price Policies, Allied Publishers Pvt. Ltd., Delhi.

5. Saxena, N.V. (1998) : What is meant by people's participation, Journal of Rural Development, Vol. 17, No.1, National Institute of Rural Development, Hyderabad.

15

Use of Chemical Fertilisers in Various Districts of Andhra Pradesh—A Study

M. Raja Sekhara Bapu
B. Sambasiva Rao

Fertilisers are essential inputs for boosting the output of various crops. Every tonne of fertiliser increased foodgrain yield by 10 to 12 tonnes. Fertilisers provide additional nitrogen, phosphorus and potash to the crops. Their use in optional quantities is essential for maintaining the fertility of land to obtain the benefit of high yielding variety seeds. Because of limited availability of land, their use becomes all the more important not only for increased yield for higher income for the farmers but also to meet the basic food requirements of the giving population.

Scenario in 1998-99

The consumption of fertilisers in the country rose to 163.1 lakh tonnes in 1997-98 (fiscal) from 143.1 lakh tonnes in 1996-97. In 1998-99, however, the consumption was expected to be lower than in 1997-98 due to lower production, lower imports and lower off-take by farmers due to poor weather conditions, i.e. floods and unseasonal rains in some reasons and dry spell in some areas which affected the overall output of many agricultural commodities including foodgrains. Broadly, about 50-55 per cent of the fertiliser in the country is used in the cultivation of paddy, wheat and sugarcane; 10-15 per cent in cotton and groundnut; 15 per cent in fruits and vegetables; 10 per cent in coarse cereals; and 10 per cent in oilseeds and pulses. Although India is world's third largest producer and consumer of fertilisers, its average consumption of fertilisers is only 77 kg against 345 kg per hectare in China. Table 1 provides the statewise consumption of all fertilisers. This paper attempts to study:

1. Need for the use of chemical fertilisers.
2. District-wise consumption of chemical fertilisers in Andhra Pradesh.

In India, there is a shortage of organic carbon in soil because of poor use of organic manure. This has reduced the efficiency of soil to absorb fertilisers and give better yields. The organic carbon is basically produced by humus or decomposed residue derived from hay and straw left after each harvest. In India, all the hay and straw is fed to animals as fodder while in other countries a part of it is allowed to remain in the soil where it is decomposed by the combined action of water and sun rays. On the other hand, in the past 15 years, major changes in the cropping pattern have emerged in Madhya Pradesh, Maharashtra, Rajasthan and Gujarat etc., which has changed the profile of fertiliser consumption in these states. The production of soyabean, sunflower, castor, mustard and fruits has gone up in these states considerably.

The use of fertilisers is increasing along with the high yielding varieties. In 1980-81, the consumption of fertilisers is 5.75 lakh tonnes is increased to 15.82 lakh tonnes in 1991-92. This shows the growing importance of fertilisers in Agricultural sector. "Rythu Kalimi, Chenuku Balimi" is the proverbs of Telugu people. The farmers in Andhra Pradesh use fertilisers according to the economic stability

Table 1: State-wise Consumption of All Fertilizers (Kgs/hectare)

State	*1988-89*	*1989-90*	*1990-91*	*1991-92*	*1992-93*	*1995-96*	*1996-97*
Andhra Pradesh	103.06	115.58	122.17	119.35	114.20	136.99	138.37
Bihar	57.82	54.14	57.44	57.94	57.18	72.95	79.61
Gujarat	61.97	71.07	72.20	74.94	73.24	67.35	72.72
Haryana	84.69	94.78	103.77	112.77	107.79	121.38	127.70
Karnataka	67.68	64.31	68.76	74.76	64.40	77.97	68.75
Madhya Pradesh	30.13	30.05	36.19	36.07	35.33	33.47	39.42
Maharasthra	45.42	54.44	60.67	58.22	52.14	64.15	62.05
Orissa	17.63	22.05	20.75	21.11	21.96	24.43	25.79
Punjab	151.00	154.86	162.02	168.36	162.23	164.22	156.99
Tamil Nadu	119.14	114.65	121.80	123.01	117.19	107.51	112.58
Uttar Pradesh	84.58	82.53	88.41	88p.72	86.00	101.22	107.57
West Bengal	76.47	60.37	90.18	90.48	87.58	97.24	102.79
All India	61.30	63.86	69.26	70.27	67.09	73.75	76.05

and the crops that they grow. Table 2 analyses about districts in Andhra Pradesh which use more fertilisers per hectare is Krishna (252.26 kgs), West Godavari (244.08 kgs), Ranga Reddy (241.11 kgs), Warangal (237.45 kgs). The lowest consumption of fertilisers is found in the districts of Adilabad (48.18 kgs), Anantapur (47.01 kgs), Visakhapatnam (57.82 kgs). The region-wise analysis depicts that the average use of fertilisers per hectare in coastal districts is 176.32 kgs, in Telangana 152.19 kgs and in Rayalaseema 88.85 kgs.

Table 2: District-wise Consumption of Chemical Fertilisers in AP (1998-99)

District	*Area (L.Hec)*	*Usage of Chemical Fertilisers (Lgs/Hec)*
1	2	3
Srikakulam	4.25	92.37
Vijayanagaram	4.58	65.74
Vishakhapatam	4.50	57.82
East Godawari	7.82	170.66
West Godawari	6.88	244.08
Krishna	7.39	252.26
Guntur	8.71	207.39
Prakasam	5.95	167.67
Nelore	3.61	230.14
Coastal	53.69	1488.13
Chittoor	5.41	86.50
Cuddapah	4.57	122.66
Ananthapur	10.65	47.01
Kurnool	10.19	118.65
Rayala Seema	30.82	374.82
Mahaboobnagar	8.64	94.98
Rangareddy	3.43	241.11

(Contd...)

1	2	3
Medak	3.88	138.08
Nizamabad	5.78	154.23
Adilabad	5.80	48.18
Karimnagar	6.49	179.73
Warangal	5.05	237.45
Khammam	7.24	109.88
Nalgonda		
Telangana	46.31	1203.64
Andhra Pradesh	130.82	147.37

Source: Director of Agriculture, Andhra Pradesh

Table—3 provides the information about the difference between the suggested dose and actual used dose of fertilisers to various crops in the Guntur district. Out of 6 lakh hectares of cultivable land in Guntur district where chemical fertilisers are used heavily 4 lakh tonnes of

Table 3: Difference Between the Suggested Dose and Actual Used Dose of Fertilisers to Various Crops in the Guntur District

Sl. No.	*Crop*	*Suggested dose in bags (50 kgs) per acre*	*Used/consumed dose in bags (50 kgs) per acre*
1.	Paddy	3	5
2.	Cotton	4	8
3.	Chillies	5	10
4.	Turmeric	6	10
5.	Banana	11	20
6.	Vegetables	4	6
7.	Pulses	1	1

Source: "Pasidipanatalu" Agricultural Magazine (Telugu), dt. 15.8.2000, p. 1.

chemical fertilisers are used that means on an average 660 kgs per hectare, and 265 kgs per acre or 5 bags of fertilisers are need. For banana crop, the suggested dose of fertiliser is 11 bags but the used dose is 20 bags, i.e., double of the suggested dose. The reason for this type of usage is the subsidy that they are getting from the government. That indicates the increase of fertilisers in Agricultural crops.

We can conclude that even though the prominence of fertilisers has been increased considerably from 1950, but from 1970 the growth has been increased enormously along with the use of high yielding varieties and from 1980 the consumption was increased unscrupulously. The following are the suggestions forwarded for implementation:

Suggestions

1. Reduce the indiscriminate use of fertilisers.
2. The system of crop rotation/crop diversity should be used.
3. Farm yard manure should be used as it is like mother's milk, because it contains all the nutrients required for all crops.
4. Regulation of fertiliser distribution should be made so that small and marginal farmers will get use the required amount of fertilisers.
5. Instead of giving subsidy on fertilisers better give subsidy on cultivation.

REFERENCES

1. Vikas Singhal (1999), Indian Agriculture.
2. Handbook of Statistics (1999), Andhra Pradesh.
3. Handbook of Statistics (1999), Guntur.
4. Ch. Hanumantha Rao (1975), Technological Change and Distribution of Gains in Indian Agriculture.
5. Government of India: Ministry of Finance, Economic Survey, 1997-98.

16

Mechanisation and Labour Absorption in Andhra Pradesh Agriculture–A Study in the Right Canal Commond Area of NSP

L.K. Mohana Rao
P. Narasimha Rao

Since the dawn of the industrial revolution mechanization has become an irreversible process, subsuming developed and developing economics, with a spurt in technological innovations during the last forty years, there has been a phenomenal acceleration of mechanization in the less developed countries. In the current era of progressive economic globalisation and transnationalisation of capital, production systems and markets are often compelled to respond to a single global logic of exchange. This process, geared to unlimited competition, puts a premium on consideration of efficiency even in the operations of remote farms which were once islands of subsistence agriculture for small and marginal farmers or pockets of random neglect of the absentee landlord. In developing economies like India, where unemployment or under-employment is a serious problem, divergent views are expressed with regard to the mechanization of farm operation in agriculture. One set of writers (Ashok Mitra and others) who are somewhat ideologically inclined, posit the view that mechanization has deleterious effects, as it is perceived to result in reduction of employment. The other set of studies are empirically oriented and based on field work. Among them, the findings of Bose and Clark, Billings and Singh, Grewal and Kahlon, Suzuki, Sison, etc., suggest that mechanization leads to reduction in employment. On the other hand the findings of Raj Krishna, Rao, Wolf, Chopra, Jhol, etc., are categorical that mechanization does not result in reduction in employment. Keeping this in view, an attempt has been made in this study to examine that impact of mechanization on human labour employment in Andhra Pradesh where mechanization particularly use of combined harvester in agriculture is going at rapid rate.

Objectives

The objective of the study is to examine the impact of mechanization on human labour absorption in agriculture in Andhra Pradesh.

Methodology

The data for this study is drawn from an ongoing Research Project 'Equity in Irrigation. A study in the Right Canal Command Area of Nagarjuna Sagar Project' sponsored by U.G.C., New Delhi. The sample comprises of 230 farm households and the distribution of sample holdings among different categories are presented in Table 1. Decomposition analysis is carried out to capture the changes in human labour employment caused by different factors. The analysis is carried out at the aggregate level (all crops put together) and at the disaggregate level (for two predominant crops paddy and cotton). The final equation adopted to decompose the changes in human labour employment into different components is associated with employment with respect to land, fertilisers, pesticides and capital services respectively on mechanised farms.

Table 1: Distribution of Selected Holdings

Category of Farm	*Mechanised*	*Non-mechanised*	*Total*
Marginal	32 (31.07)	50 (39.37)	82 (35.65)
Small	33 (32.04)	38 (29.92)	71 (30.87)
Medium	25 (24.27)	24 (18.90)	49 (21.30)
Large	13 (12.62)	15 (11.87)	28 (12.17)
Total	**103** **(100.00)**	**127** **(100.00)**	**230** **(100.00)**

(Figure in parenthes is are percentages to the total).

$1_n [X_{M1}/X_{n1}] = \{1_n (A_M/A_n)] + [(A_2 - B_2) 1_n X_{N2} + (A_3 -]$

$B_3) 1_n X_{N3} + (A_4 - B_4) 1_n X_{N4} + (A_5 - B_5) 1_n X_{N5}]$

$+ [A_2 1_n (X_{m2} / X_{N2}) + A_3 1_n (X_{M3} / X_{N3}) + A_4 1_n (X_{M4}/ X_{N4}) +$

$A_5 1_n (X_{M5} / X_{N5})]$

In the above, the suffix 'M' and "N' denotes for mechanised and non-mechanised farms respectively.

X_1 : human labour employment per acre of cropped area

X_2 : size of the land in acres

X_3 : value of fertilisers and manures per acre

X_4 : value of pesticides per acre

X_5 : value of capital services per acre

A_m and A_N are the intercept terms

A_2, A_3, A_4 and A_5 are the elasticity coefficients

B_2, B_3, B_4, and B_5 are the elasticity coefficients associated with employment with respect to land, fertilisers, pesticides and capital services respectively on non-mechanised farms.

In the equation used for decomposition analysis, the bracketed expression on the left-hand side is the natural logarithm of ratio of per acre labour used on mechanised and non-mechanised farms and reflect the approximate total percentage change in per acre labour employment due to mechanisation. The first two bracketed expressions on the right hand side of the equation measures the changes in employment per acre due to change in the intercept term and employment elasticities. The second bracketed expression calculates the sum of changes in the employment elasticities weighted by the logarithm of farm size, value of fertilizers, pesticides and capital services used per acre under non-mechanised conditions. Therefore, the first two bracketed expressions together gives the percentage change in the human labour demand exclusively due to mechanisation or represent the net impact of mechanisation on human labour demand. The third bracketed expression is the sum of the logarithm of the ratio of the farm size and of the ratios of fertilizers, pesticides and capital services under mechanised and non-mechanised conditions. This measures the percentage change in human labour demand due to change in farm size, and changes in the use of fertilizers, pesticides and capital services, given the employment elasticity of these inputs on mechanised farms.

Results

The estimated results of the decomposition exercise are presented in Table 2 for all crops, paddy and cotton. The results reveal that there is close association between the observed change and the estimated change in the employment in all the crop situations. This may imply that the decomposition equation specified for this purpose is robust.

All Crops

As all the crops put together, the labour use per acre on mechanised farms is found to be higher by 2.56 per cent as compared to non-mechanised farms. The decomposition analysis suggests a decline of 13.41 per cent (Table 2) in the use of human labour employment with the

Table 2: Decomposition Analysis Results of Total Difference in Human Labour Employment—All Crops

Item	*Percentage attributable*			
	All Crops	*Marginal*	*Small*	*Marginal and Large*
Total observed difference	2.56	2.54	5.60	0.50
1. Mechanisation	–13.41	–14.00	–4.67	–11.51
2. Changes in inputs:	15.92	16.34	10.30	11.89
(a) Land	–1.76	2.71	–0.05	–4.55
(b) Fertilizer	8.18	14.91	8.58	8.34
(c) Pesticides	8.18	1.98	2.70	5.22
(d) Capital Services	1.32	–3.26	–0.93	2.88
Total estimated change due to all sources	2.51	2.34	5.63	0.38

adoption of mechanisation. The difference in the farm size between mechanised and non-mechanised farms has resulted a decline in human labour employment by 1.76 per cent given the elasticity of employment with respect to farm size of non-mechanised farms. On the other hand, given the elasticity of employment with respect to fertilizers, pesticides and capital services of non-mechanised farms, the difference in the level of use of these inputs has resulted an increase in the human labour employment by 17.68 per cent. This may imply that the negative effects of mechanisation on employment is outweighed by the positive effects of fertilisers, pesticides and capital services inputs. A similar pattern with variations in the relative impact of different factors on employment is descernable among different size groups of farms. The total estimated increase in human labour employment with the adoption of mechanisation on marginal, small and large farm categories worked out to 2.34 per cent, 5.63 per cent and 0.38 per cent respectively. This may mean that the overall effect of mechanisation on human labour employment is positive and higher on small and marginal farms as compared to large farms.

Paddy

With regard to paddy crop, the labour use per acre on non-mechanised farms is higher by 1.57 per cent than mechanised farms. However, different pattern is found among different size groups (Table 3). The average labour use per acre on marginal and small farms of mechanised

Table 3 Decomposition Analysis Results of Total Difference in Human Labour Employment-Paddy

Item	*Percentage Attributable*			
	All Crop	*Marginal*	*Small*	*Marginal and Large*
Total observed difference	–1.57	2.54	8.82	–6.18
1. Mechanisation	–13.29	–14.00	–6.76	–13.75
2. Changes in inputs	11.79	16.34	15.61	7.98
(a) Land	–2.70	2.71	–6.47	–10.04
(b) Fertilizer	10.02	14.91	17.11	8.78
(c) Pesticides	4.42	1.98	2.38	5.33
(d) Capital Services	0.05	–3.26	2.59	3.91
Total estimated change due to all sources	–1.50	2.34	8.85	–5.77

category is higher by 2.54 per cent and 8.82 per cent respectively as compared to the corresponding size groups of non-mechanised category. On the other hand, on medium and large farms, the per acre labour use on non-mechanised category is higher by 6.18 per cent than machanised category.

The results of the decomposition analysis reveal that, mechanisation alone has reduced labour employment by 13.29 per cent given the input use levels of non-mechanised farms. The farm size differential between mechanised and non-mechanised farms has resulted a decline in human labour by 2.70 per cent given the elasticity of human labour with respect to land on mechanised farms. On the other hand, given the elasticities associated with fertilisers and pesticides on mechanised farms, these inputs together have increased human labour employment by 14.44 per cent. The capital services input has also increased the human labour employment on mechanised farms marginally.

A similar pattern, with few exceptions, is observed among different size groups with respect to the change in employment caused by mechanisation and changes in the use of other inputs. Here also the difference between observed and estimated change in employment is found to be marginal.

Cotton

Unlike in paddy, cotton under mechanised conditions has absorbed 7.34 per cent higher human labour per acre compared to the human labour employment under non-mechanised conditions (Table 4). The estimated results of the decomposition analysis indicate that mechanisation in cotton cultivation has declined the human labour employment by 15.01 per cent given the input use level of non-mechanised farms. The differences in the use of other inputs, viz, land, fertilisers and pesticides has resulted an increase in the human labour employment on mechanised holdings by 1.28

Table 4: Decomposition Analysis Results of Total Difference in Human Labour Employment–Cotton

Item	*All Crop*	*Percentage Attributable* *Marginal*	*Small*	*Marginal and Large*
Total observed difference	7.34	13.08	20.45	2.62
1. Mechanisation	–15.01	6.59	–7.70	–3.50
2. Changes in inputs:	21.76	6.48	26.71	5.91
(a) Land	1.28	–3.63	–0.83	–5.62
(b) Fertilizer	10.42	1.41	15.08	6.14
(c) Pesticides	10.10	7.29	10.90	5.63
(d) Capital Services	–0.04	1.41	1.56	–0.24
Total estimated change due to all sources	6.75	13.07	19.01	2.41

per cent, 10.42 per cent and 10.10 per cent respectively. The total estimated increase in labour employment per acre in cotton production from all sources is 6.75 per cent. The same pattern prevails among different size groups of farms also.

Conclusions

From the above discussion it can be inferred that in the overall crop production as well as in the production of paddy and cotton, mechanisation has negative effects on human labour employment in a direct manner. However, in all the crop situations the impact of other inputs, specifically fertilisers and pesticides, on labour employment is positive. This may mean that mechanisation by way of promoting intensive cultivation may lead to the intensive use of farm inputs and thereby enhancing human labour employment. Consequent to this, the overall effect of mechanisation and other farm inputs put together on human labour employment is positive.

REFERENCES

Abercrombie, K.C. (1973), "Agricultural Mechanisation and Employment in Latin America," Mechanisation and Employment in Agriculture: Case Studies from Four Countries, International Labour Office, Geneva.

Anderson, *et. al.* (Ed) (1982), Science, Politics and Agricultural Revolution in Asia, Westview Press, Colarado.

Bergman, Theodor (1978), Mechanisation of Indian Farming, Popular Prakashan, Bombay.

Bhatia, Bhimsen (1990), Adoption of Farm Mechanization in a Developing Economy, Daya Publishing House, New Delhi.

Bhattacharya, N. and Saini, G.R. (1970), "Mechanization and Rural Employment with some implications for Rural Income Distribution," Economic and Political Weekly, RA, Vol.5, No.26.

Billings, M.H. and Singh, A. (1970), "Mechanization and the Wheat Revolution Effects of Female Labour in Punjab", Economic and Political Weekly, Vol. 5, No. 52.

BOSE, S.R. and Clark, E.H. (1969), "Some basic considerations on Agricultural Mechanisation in West Pakistan", The Pakistan Development Review, Vol. 9, No. 3.

Clayton, E.S. (1974), "Farm Mechanisation and Employment in Developing Countries", International Labour Review, Vol. 110, No. 1.

Grewal, S.S. and Kahlon, A.S. (1972), "Impact of Farm Mechanisation on Farm Employment in Punjab, Indian Journal of Agricultural Economics, Vol. 27, No. 4.

Johl, S.S. (1970), Mechanisation, Labour Use and Productivity in Indian Agriculture, Economics and Sociology Occasional Paper No. 23, Ohio State University, Columbus, Ohio.

Johnston, Bruce F. and John Cowine (1969), "The Seed Fertilizer Revolution and Labour Force Absorption," American Economic Review, Vol. 59, No. 4.

NCAER (1973), Impact of Mechanization in Agriculture on Employment, National Council of Applied Economic Research, New Delhi.

Stewart, Frances (1972), "Choice of Techniques in Developing Countries", Journal of Development Studies, Vol. 9, No. 1.

Suzuki, F. (1983), Farm Mechanization in Asia, Asian Publishing Organisation, Tokyo.

17

Food Grain Production in Andhra Pradesh

I. Satya Sundaram

Andhra Pradesh, know as 'Rice Bowl of India' is facing disturbing developments in the agricultural sector.

In Andhra Pradesh, the primary sector's contribution to total state domestic product declined from 43 per cent in 1980-81 to 33 per cent in 1990-91, and further to 29 per cent in 1998-99. Yet, the labour force dependent on agriculture as a proportion of total labour shows only slight decline from 69.5 per cent in 1980-81 to 68.6 per cent in 1990-91.

Andhra Pradesh is favourable placed in terms of factor endowments. It has a higher irrigation ration, and per hectare fertilizer and labour inputs. Yet, the state has not been able to achieve results compatible with these favourable factors[1].

Growth and Performance

The state's agricultural sector experienced a growth rate of only two per cent in the 1980s and 1990s. Net irrigated area has been almost stagnant, in the 1990s.

In respect of rice yields, Andhra Pradesh ranks fourth behind Punjab, Tamil Nadu and Haryana, with 2494 kgs per hectare. Going by the productivity, it is ranked second in maize.

An important factor contributing to decline in food grain production in the state is the continuing shift of area from food grains to non-foodgrains[2].

For the state as whole, the area under food crops is 68.1 per cent, while 31.8 per cent area is under non-food crops.

Gross cropped area in the state increased from 121.01 lakh hectares in 1985-86 to 130.43 lakh hectares in 1995-96.

The area under rice increased from 29.61 lakh hectares in 1960-61 to 32.07 lakh hectares in 1987-88 while the yield increased from 1238 kg/ha to 2258 kg/ha.

Paddy cultivation alone accounts for 65 to 75 per cent of the total irrigation potential.

Andhra Pradesh occupies third position in rice production in the country after West Bengal, and Uttar Pradesh, accounting for nearly 12 per cent of the country's total rice production. It is the Second major supplier of rice to the buffer stocks of the country, after Punjab.

It is estimated that more than 28 per cent of the gross cropped area and 50 per cent of the area under food crops in the state is under rice cultivation.

The quality of fertilizer use had increased substantially over the last decade.

However, the farm level use of fertilizer was about 60 per cent higher than the recommended levels for intensive rice systems in both canal irrigated rice region and non-canal irrigated rice region. This may be due to the government subsidy on urea. However, higher uses of fertilizers in leading to heavy pressures on soil fertility.

Food grain Production in Andhra Pradesh (in thousand tonnes)

Year	*Total Cereals*	*Total Pulses*	*Total Food Grains*
1990-91	11634.1	695.5	12329.6
1991-92	10913.6	791.8	11705.4
1992-93	10918.9	739.0	11657.9
1993-94	11576.5	677.0	12253.5
1994-95	11111.1	672.5	11783.6
1995-96	10895.4	771.2	11666.6
1996-97	12837.2	838.0	13675.2
1997-98*	10306.6	515.7	10822.3
1998-99	16351.9	763.4	14395.3

Source: Economic Survey 1999-2000, p. 8-19.

Major Problems

In Andhra Pradesh, farmers are experiencing considerable yield losses in paddy from various biotic and abiotc pressures. It is estimated that the yield loss is about 530 kg per hectare of which 60 per cent is due to biotic stresses (insects and disease) and the remaining due to pressures from resource degradation (soil and water). The total loss of paddy during 1990-2000 period is estimated at 3.2 million tonnes valued at Rs. 1750 crores[3].

There is also skewed pattern of agricultural development in Andhra Pradesh. For instance, six districts (East Godavari, West Godavari, Nellore, Chittoor and Nalgonda) accounted for 75 per cent of the incremental rice production in the post HYV period[4].

It is true that the per capita gross cropped area in Andhra Pradesh in 1991 was slightly higher than the national average, that is, 0.23 ha as against 0.21 ha. But, the per capita income generated from agriculture (Rs. 1240) in the state was lower than the national average (Rs. 1522).[5]

One reason for this is the regional bias in the additional irrigation potential created in the post-independence period[6].

Water is a crucial input, the availability or other wise of which will determine to a large extent the crops raised, their yields and their stabilisaiton of returns[7].

Heavy investments have been made in the irrigation sector during the plan periods. Yet, 60 per cent of the cultivable land in the state is rain-fed. Added to the poor utilization of irrigation potential are the frequent occurrence of natural calamities such as droughts and cyclones.

The percentage of irrigated are a to gross cropped area has gone up to 43.1 per cent only from 27.2 per cent in 1960-61. At the national level, during this period, it has more than doubled to 37.5 per cent from 17.2 per cent.

According to Prof. G. Parthasarathy, the fluctuations in agricultural production in the state is due to the excessive dependence on tank irrigation. Most tanks are silted. Their storage capacity is coming down[8].

It is felt that minor irrigation which is cost effective and whose spread effects are strong has been neglected.

In recent times, the total area under irrigation has declined from 5.5 million hectares in 1989-90 to 5.0 million hectares in 1993-94. This is due to the decline in the area irrigated by public surface network. This has serious implications.

What is worse the gap between the potential created and utilized under surface irrigation is about 25 per cent. This indicates significant inefficiency in the surface irrigation network[10].

As in most states, in Andhra Pradesh, small and marginal farmers account for 80 per cent of farming households, and 40 per cent of area cultivated. The proportion of agricultural labour households in the state has been on the increase, and it is the highest in the country–about 40 per cent of the workforce.

Measures Needed

Market approach though has its merits is not the answer for ills, as markets pay scant attention to distributive justice and ecology. Neglect of participatory development would adversely affect agriculture[11].

The state government has adopted a three-pronged strategy to exploit the vast unutilized potential in the agricultural sector. This Includes:[12]

- *(i)* fully harnessing the water resources by completing all the pending irrigation projects and planning and executing new projects in the shortest possible time;
- *(ii)* focusing on rain-fed agriculture through revolutionary schemes like watershed management and drip-irrigation; and
- *(iii)* spurring the growth of agriculture through implementation of radical reforms in the sector.

The state government has devised a new approach, which is crop-specific. It chalked out a detailed farming situation based plans district-wise and specific schemes are proposed to fill the gaps in productivity in each situation[13].

To make the best use of scarce water and land resources in drought-prone areas, it is necessary to introduce certain collective or cooperative action at least with 'watershed' as a unit of organisation[14].

Thus, food grain production in Andhra Pradesh needs to be stepped up by a judicious use of crucial inputs like water and fertilizers. The state's potential has to be fully utilized.

REFERENCES

1. R. Radhakrishna: 'Andhra: Rich in Factor Endowments'. The Economic Times, February 28, 1991.
2. B. Parthasarathy and K.A. Nirmala: 'Andhra Pradesh Agriculture: New Challenges', The Economic Times, November 1, 1999.
3. The Hindu Business Line, June 9, 2000.
4. S. Subrahmnanyam and P. Satya Sekhar (1990): 'Rice Economy' in K. Mukund (Ed): Andhra Pradesh Economy in Transition. Book Links Corporation, Hyderabad.
5. R. Radhakrishna. *op. cit.*
6. M.V. Srinivasa Gowda, M. Rani and A. Sachindra Babu: 'Economic Development of Andhra Pradesh: A Comparative Assessment' Southern Economist, December 1 and 15, 1997, p. 8.
7. K. Chakradhar Rao: 'Nature and Impact of Technological Changes in Andhra Pradesh Agriculture', Southern Economist, March 15, 1989, p. 15.
8. Interview with Prof. G. Parthasarathy. The Economic Times, October 22, 1995.
9. R. Radhakrishna: *op. cit.*
10. The Economic Times, April 17, 1997.
11. V. Ratna Reddy: 'Farm Sector in AP: Poised for Major Changes', The Hindu, July 11, 2000.
12. The Economic Times, November 1, 2000.
13. *Ibid.*
14. G. Parthasarathy: 'Agriculture in AP', The Economic Times, October 1, 1985, p. 17.

18

An Analysis of Pulses Production in Andhra Pradesh

M.V. Narasimha Sarma
G.B. Franklin
T. Hanumantha Rao

Pulses are protein rich food products, which are very essential for balanced diet. Statistics reveal that around 13 per cent of agricultural land is under the cultivation of pulses in Andhra Pradesh of which only 11.7 per cent of land has got irrigation facility. In term of area, under cultivation of pulses in Andhra Pradesh, Guntur district stood first with 20.2 per cent followed by Srikakulam (18.9), East Godavari (17.1), Krishna (16.4), Prakasam (15.3), Visakhapatnam (15.2) and West Godavari stood last with 1.5 per cent.

Scenario of Pulses Production in India and Andhra Pradesh

Stability in production of pulses mainly depends on the area under cultivation. The average production in Andhra Pradesh is less than the national average. The following table shows the production of pulses in some selected states in India. Eventhough there was an increase in area under cultivation and use of fertilizers, the production of pulses has not increased in accordance to their increase. Table—1 shows the annual output of Red gram per hectare in selected states in India.

Table 1: Table Showing the Annual Yield of Redgram in Selected States in India

Sl. No.	*State*	*Average yield Per hectare (In quintals)*
1.	Andhra Pradesh	0.34
2.	Bihar	1.07
3.	Uttar Pradesh	1.05
4.	Tamil Nadu	0.77
	INDIA	0.66

Table 2 shows the annual output of Red gram per hectare in selected states in India and Andhra Pradesh. Compared to other states given in the table, average yield per hectare of Andhra Pradesh (0.34) is less than national annual average (0.66)

Table 2: Table Showing the Annual Yield of Pulses Excluding Red Gram in Selected States in India

Sl. No.	*State*	*Average yield Per hectare (In quintals)*
1.	Andhra Pradesh	0.48
2.	Uttar Pradesh	0.79
3.	Kerala	0.71
4.	Haryana	0.64
	INDIA	0.45

Above table shows that average annual yield of Andhra Pradesh slightly above the national yield in other pulses production.

Objectives

The specific objectives of this paper are:

1. To find out the trend equations for output and area under cultivation for different pulses during the study period.
2. To find out the relationship between the prices and area under cultivation of pulses (supply effect).
3. To suggest different measures for the better performance of pulses cultivation in the wake of liberalization.

Data and Methodology

This paper is based on secondary data collected by browsing from internet[1] on area, out turn and average wholesale prices of pulses in Andhra Pradesh complied by Directorate of Economics and Statistics, Andhra Pradesh. In order to study the objectives set forth, percentages, averages, method of least squares to find out the trend and linear regression equations to find out the relationship between prices and area under cultivation are used.

An Analysis of Pulses Production in Andhra Pradesh

In accordance to the objectives set above an attempt has been made to find and the trends in area and output for different pulses crops by using least squares equation.

The area shows a negative increase in case of Bengalgram (–1.08), Greengram (–7.11), Blackgram (–11.11), Horsegram (–6.02) out other pulses (–0.10). Even-though there is a substantial positive increase (18.09) in the area sown for redgram, it could not over take the negative growth (7.28) in the total area of pulses cultivation.

Coming to the trend in the output produced, there are some interesting results to note. In case of Bengalgram,

Blackgram, and Horsegram there is a negative growth in output, with –2.38, –5.39 and –1.75 grown rates, which is in line with negative growth of area sown. But in case of redgram, greengram, other pulses and total pulses, even though, the area sown shows a negative growth, the output has a positive increase. The main reason for such increase in output is due to better use of irrigation facilities and increase in the consumption of fertilizers and pesticides.

Table 3: Table Showing Regression Equations for Area and Out-turn of Different Pulses

Sl. No.	*Name of the Crop*	*Least Square Equation*	
		Area	*Out-turn*
1.	Bengalgram	139.09–01..08t	103.95–2.48t
2.	Redgram	258.75+18.09t	83.67+9.87t
3.	Greengram	497.58–07.11t	173.28+4.76t
4.	Blackgram	526.52–11.11t	291.19–5.39t
5.	Horsegram	138.52–06.02t	46.68–1.75t
6.	Others	35.30–0.10t	8.73+0.29t
7.	Total	166.48–7.28t	710.93+5.21t

The supply response to price for any commodity is well-established fact. As far as agriculture is concerned the current years production is always related to the price in the previous year. As such a study has been done with the following regression equation based on the following function:

$Y = f(P_{t-1})$

Where Y = Output of a pulse

$P_{(t-1)}$ = Price of the pulse in the previous years

Table 4: Table Showing Regressions Equations of Y on $P_{(t-1)}$

Sl.No.	*Name of the Crop*	*Regression equation of Y on $P_{(t-1)}$*
1.	Bengalgram	$17.55+5.68\ X_{(t-1}$
2.	Redgram	$92.36+0.13\ P_{(t-1)}$
3.	Greengram	$193.32-1.09\ X_{(t-1)}$
4.	Blackgram	$101.80+0.25\ P_{(t-1)}$
5.	Horsegram	$60.00-2.73\ P_{(t-1)}$

The supply price response is positive in case of Bengalgram, Redgram and Blackgram. However, the response is negative is case of Greengram and Horsegram. Basically, on this observation it may be concluded that the supply response effects production of pulses.

Suggestion

In the recent days farmers in Andhra Pradesh are facing a threat in the wake of Globalization sand liberalization for selling their produce. The cereal producers in Andhra Pradesh are affected much due to the drastic fallen of prices of their produce in the recent past.

In light of this and the observations made above it maybe suggested that the farmers should shift from the production of cereals to pulses for their well-being and to face the international threat.

The study also suggests a positive supply response for pulses in Andhra Pradesh. This will make in to suggest for conducting awareness Programme as a past of Janmabhoomi for production of pulses should be taken up in Andhra Pradesh to give a solace and to boost up the confidence on farming to the farmers of Andhra Pradesh.

19

Agriculture in Andhra Pradesh: Current Trends and Propspects for Future

C.S.N. Raju
K. Raja Babu

Introduction

Andhra Pradesh is known world over as traditionally agricultural state providing employment to 65 per cent of Population of the state while 30 per cent of GSDP is contributed by agriculture and allied sectors. Andhra Pradesh is unique in having seven different agro-climatic regions. Some regions are dry and bereft or irrigation facilities where as some are wet and humid frequently attacked by cyclonic storms. Despite all these problems the state farmers are known for their tenacity and ability to overcome adverse conditions to come out with bumper harvest year after year. Recently Government of Andhra Pradesh has issued a strategy paper on agriculture. This paper seeks to highlight some of the important strategies of Government and discuss the relevance in the context of recent changes in agriculture sector.

Current trends in Agriculture in Andhra Pradesh

The following table shows the performance of Andhra Pradesh in agriculture ansd allied activities in India:

Table 1:

S. No.	Item	Rank in India	Quantity		Year
1.	Food grain production	4	14.90	Mill tons	1998-99
2.	Rice	3	11.87	Mill tons	1998-99
3.	Rice Productivity	3	2812	kg/ha	1998-99
4.	Maize	2	1.30	Mill tons	1998-99
5.	Maize Productivity	1	3328	Kg/ha	1998-99
6.	Oil Seeds	4	2.26	Mill tons	1998-99
7.	Ground Nuts	3	1.92	Mill tons	1998-99
8.	Sun Flower	3	0.24	Mill tons	1998-99
9.	Cotton	3	1.54	Mill tons	1998-99
10.	Tobacco	1	0.25	Mill tons	1998-99
11.	Sugar Cane	5	16.70	Mill tons	1998-99
12.	Mango	1	3.00	Mill tons	1998-99
13.	Coconut	3	19.20	Mill tons	1998-99
14.	Chillies	1	0.50	Mill tons	1998-99
15.	Marine Fish	1	1.50	Mill tons	1998-99
16.	Fresh Water Fish	2	2.61	Mill tons	1998-99
17.	Milk	7	4.50	Mill tons	1998-99
18.	Buffalo and Sheep	3	2.00	Mill tons	1998-99
19.	Egg	1	575.00	Mill tons	1998-99
20.	Mulberry Cultivation	2	1.20	Mill tons	1998-99
21.	Cocoon	2	35.00	Mill tons	1998-99

Source: Compiled from 'Strategy Paper of Agriculture and Allied Departments', Government of Andhra Pradesh, Hyderabad, 2000.

Table 1 amply shows that Andhra Pradesh has made significant progress in several areas of agriculture and allied sectors. Nevertheless, the government seeks to intensify agriculture production, eradicate poverty and promote employment opportunities. To achieve these goals the following strategies are contemplated.

Agricultural Development: Strategies and Targets

It is contemplated that in the next two decades, food grain production should reach a target of 20 million tonnes with rice production to reach 16 million, tonnes. In addition to rice, production of millets, pulses, oil seeds and commercial crops are also expected to reach higher levels. Further, the state is expected to focus more attention to horticulture, dairy, poultry, fisheries and sericulture.

Strategies

Agriculture Production: In order to increase agricultural production, the strategies are as follows:

1. Increase in poductivity
2. Cost reduction
3. Improvements in extension programmes
4. Strengthening the market facilities
5. Value addition
6. Quality control
7. Diversification

Increase in Poductivity

In pursuit of increase in productivity, the government seeks to execute action plans to narrow yield gap and focus on rainfed agriculture. Already watershed approach is being implemented and mechanization is on the focus. The government has already distributed 4000 quintals is of green manure seeds and integrated nutrient management is being focused through expanded solid testing. Further through creation of WUA, improved water use efficiency is augmented. In the area of agricultural finance, Rs. 6019.32 core farm credit is being mobilized.

In the years to come, the government aspires to consolidate land holdings at the farmer level through incentives. A system of custom hiring centers for agro machinery is also being considered. At least one agricultural research center in each district is contemplated as nodal center for all research activities. Now onwards soil testing facilities will also be offered by private sector.

Reduction in Cost of Production

It is expected that the farmers will be able to reduce costs of production as integrated nutrient management is being considered. Also 358 FFS/IPM schools are run during Kharif season and 716 during Rabi. To mitigate disasters, six automatic weather forecasting stations are being established in research stations. The Government seeks to promote agricultural machinery and production of implements on large scale.

Improvement in Extension Programmes

During the last year 19,022 Rythu Mitras were formed and another 7,500 will be initiated. Emphasis on primary education is highlighted by establishment of 8,000 blackboards in addition to the already existing 15,862 in backward villages. Large number of farmers will be trained in AMCs and FTCs. Further OFED is being organized in both the seasons. More NGOs will be involved in extension programmes to provide consultancy.

Stregthening the Markets

Rythu Bandhu Pathakam will be reformed to help farmers. More emphasis will be on Rythu Bazars and this will go hand-in-glove with removal of movement restrictions. Local officials will intervene in times of crises. To help farmers in marketing, information sourcing centers are planned. Government seeks to establish and expand storage infrastructure and create price stabilization fund.

Beside all these endeavors, State Agricultural Promotion Council and dist agricultural promotion councils are being organized to facilitate agricultural growth. With reference to quality control seed village programme and pesticides coding centers are being established. Pretesting of pesticides by agricultural research centers is also under consideration. The government seeks to diversify agriculture through crop shifts, extensive farming of redgram and maize plus focus on intensity of cropping by promoting intercropping and relay cropping. A new system of market intelligence and information will be used to neutralize mismatch between production (supply) and consumption (demand).

Andhra Pradesh Budget 2001-02 and Agriculture

The budget of Andhra Pradesh (2001-02) reflects the policy of the Government outlined in a strategy paper on agriculture. Use of irrigation water for agricultural purposes will be charged now onwards. Users' charges will be imposed on rural drinking water. This will surely increase the cost of agriculture production though it is expected that farmers will be more frugal with regard to the use of water. Allocation of Rs. 372.24 crores for agriculture is expected to yield results just as 5000 new minor irrigation projects are expected to help intensive and extensive cultivation. There is no wonder that Ichamapally and Polavaram projects got a raw deal as the Government has already formulated Medium Term Framework (MTFF).

Limitations of the Strategies

Agriculture in Andhra Pradesh has come of age. This sector is facing serious problems within end from outside. This paper considered internal problems of agriculture vis-a-vis the role of Andhra Pradesh Government. Recently the Government has released a strategy paper on agriculture and allied sectors. A number of approaches specific to production, productivity, marketing, extension etc. are considered by the paper. However this paper has not considered important issues pertaining to agriculture namely, unemployment, input prices, agricultural taxation, import of farm inputs, new agricultural products, non-farm products and migration of agricultural labourers.

Government should consider how the new strategies affect employment, inputs, imports and exports. If these issues are neglected, the new strategies may not bring about significant and sustained growth in agriculture.

Unemployment

Strategies specific to cost reduction may cause unemployment and underemployment. Diversification may also cause large-scale unemployment unless agricultural labours adapt to new production techniques of new agricultural products.

Input Prices

New high yielding variety seeds, fertilizers, pesticides and other farm inputs many actually increase cost of production. With the competition from other countries and other states, Andhra Pradesh farmer many find it difficult to reduce cost of production.

Agricultural Taxation and Subsidies

Sooner or later agriculture sector may attract more new taxes which means that agricultural products many cost higher. Similarly the subsidies may also be removed in phased manner. This will also lead to higher cost of production and therefore, prices.

Import of Farm Inputs and Farm Products and Non-farm Products

Indian agricultural may have to buy directly farm inputs from abroad or from multinational corporations in India at higher prices. These may significantly affect output prices. Competition from imported products many adversely affect the domestic farmers. Non-farm products also many face similar competition from outside.

Migration

There is extensive migration of both labour and capital from agriculture sector. In the coming years, the farm sector may face serious problems due to capital flight and migration of labour.

In the light of the issues discussed earlier, it is important that the strategy paper on agriculture and allied sectors gives cognizance to these issues and spell out a more refined, more realistic and comprehensive strategies to help sustainable agricultural development.

What Can Be Done

- Rationalization of Labour Law
- Emphasis on special crops like Basmati
- Floriculture and arboriculture may be given importance
- Satellite townships may be augmented to prevent migration.
- Indigenous renewable source of energy should be promoted
- Informal money markets should be regulated and emphasis be given to micro banking.
- Efforts should be made to empower women in agriculture sector

REFERENCES

1. Government of Andhra Pradesh, Strategy Paper of Agriculture and Allied Departments, January 2001
2. V.M. Rao and H.G. Hanumanthappa, "The Marginalization Process in Agriculture: Indicators Outlook and Policy

Implications", Economic and Political Weekly, Vol. 34, No. 52, December, 25-31, 1999

3. Alberto Valde's "Agriculture in Uruguay Round: Interests of Developing Countries", The World Bank Economic Review, Vol. 1, No. 4, September, 1987

4. The World Bank, World Development Report, Oxford University Press, New York 1996

5. Government of Andhra Pradesh, Strategy Paper on Fiscal Reforms, Finance Department, February 2001

6. Delhi Science Forum, The Indian Economy, 1999-2000, An Alternative Survey, New Delhi, November 2000.

20

Non-Performing Advances in Agricultural Sector: A Review

M. Koteswara Rao
P. Venu Gopal

Banks are derivative institutions. The health of the real sector is reflected in banking system's health. If the credit discipline is weak and state intervention in the financial sector is pronounced, there is every possibility that the banking system will become weak and often unsound. In fact, 'Financial Repression' has been the bane of many economists and reforming the system has posed enormous difficulties. At the same time, quite a few countries that adopted the 'Big Bane' approach to the financial liberalization and deregulation have come to grief.

Abundant literature has emerged on the progress of financial sector in developing countries including India. An attempt is made in this paper to focus on critical issues that the Indian Banking system is facing at present, and more particularly, the huge pile-up of Non-Performing Advances (NPAs) in Agricultural Sector for which the banks have come to be saddled with. The systematic soundness and stability of the banking system all over the world has drawn the attention of policy-makers. The issue of NPAs has come to occupy the center state in the discussions on Indian Financial reforms also.

Magnitude of NPAs

The level of NPAs in the Indian Banking sector has reached an alarming proportion. In terms of absolute amount, the aggregate NPAs for the public sector banks have crossed Rs. 50,000 crores by the end of March 2000. In relative terms, the magnitude of the problem could be seen from the fact that the NPA as percentage of total advances of all public sector banks by the end of March 1993, 1994, 1995, 1996 and 1997, was of the order of 24.2, 23.6, 19.5, 17.0 and 15.75 respectively. The immediate financial effort of such NPAs level is seen from the fact that all scheduled commercial banks had to made a substantial provision of Rs. 8043 crores by 1992-93 and Rs. 8043 crores by 1993-94 to take care of the NPAs.

NPAs in Agriculture Sector

The NPAs in the Agriculture Sector advances assumed a special significance because, Firstly, the NPAs in the Agricultural Sector lie outside the purview of the external agency such as Debt Recovery Tribunals set-up for recovery of the hard-core and large NPAs of banks. Secondly, the agricultural sector is going to be an integral part of the credit policy for the commercial banks even in the future. The policy-makers have not considered the recommendations of the Khusro Committee and the Narasimham Committee for the dilution of the agricultural sector targets to any worthwhile extent. In fact, targets have been tightened. Thirdly, agricultural sector advances are spread over the length and breadth of the country and, hence, the management of NPAs will have to be done mostly at the branch level. The geographical spread of internal as well as external factors are also leading to mounting of NPAs. The effect of the one such external factor, viz, write-off of the debts of borrowers under the Agricultural and

Rural Debt Recovery (ARDR) Scheme created an effect of shock being felt by the commercial banking sector and the honest borrowers on the other.

Distribution of Banks Credit

Table 1 presents data on the percentage distribution of outstanding credit of scheduled commercial banks according to population group which shows that agriculture receives the major portion in rural area and supresses other population group such as semi-urban and metropolitian areas.

Table 1: Population Group-Wise Bank Credit Percentage Share According to Occupation

Occupation	*Rural*	*Semi-urban*	*Urban*	*Metro*	*All-India*
Agriculture	36.5	22.2	5.7	2.5	10.7
Industry	27.3	35.9	49.5	58.4	49.2
Transport Operations	2.7	2.7	1.7	1.4	1.8
Professional and Other Services	2.2	2.9	5.4	2.9	3.3
Personal Loans	10.6	15.5	14.3	7.5	10.3
Trade	14.0	12.9	14.7	13.5	13.7
Finance	0.6	0.7	7.1	7.1	4.1
All Other	6.1	7.2	7.5	6.7	6.9
Total Bank Credit	**100.00**	**100.00**	**100.00**	**100.00**	**100.00**

Source: IBAI Bulletin, February, 2000, p. 49.

The Emergence of Overdues

The overdues in Agricultural advances have emerged over the years due to a number of factors affecting directly or indirectly, the borrower's repayment schedule (Table 2).

There are three distinct stages in the emergence of overdues, inter-linked with the official policies which gave a thrust to rural credit. As the credit supply increased because of the policy intervention, the initial reluctance and non-preparedness of the bankers showed the seeds of deliquency. The roles of the two players-the lenders and the borrowers–have also undergone dramatic changes during these stages. The details of the thee stages are furnished in the Table 2.

Table 2: The Three Stages of Emergence of Overdues

Feature	*1st stage 1969-75*	*2nd stage 1976-80*	*3rd stage 1981-2000*
The Player Lenders	Reluctant Bankers	Soft-credit Lenders	Target-Oriented Lenders
Borrowers	Uninitiated farmers	Purposeful Borrowers	Subsidy Oriented Borrowers
The Means	1975	1980	1998
Rural Branches (NO)	6616	14818	32890
Agriculture Loan Accounts (crore)	0.22	0.86	1.94
The Result Agricultural Advances (Rs. in crores)	658	2418	23813
The Residue Amount	N.A.	447*	4965#
Overdue (Rs. in Crores)			
Overdue (Percentage)		47.8*	38.1#

*June 1979 data.
June 1996 data.

Stage-1

Immediately after bank nationalisation, the thrust was on rural banking. The banks expended their branch network which increased to total branch network. They reached out to farmers and buid up 0.22 crore borrowing accounts. The advances rose to Rs. 658 crores by the end of this period. The bankers were the reluctant players in the rural scene. They were not prepared organizationally and attitudinally to serve the rural economy. When the central bank advised them to explore avenues for credit development within the villages, the banker rolled up his sleeves and sat down to finance the large farmers for purchase of tractors, power tillers and pumpsets, by making land as collateral security. Lendings were sporadic and scattered. The dealers in these farm implements collaberated with the branch managers, in many cases, for financing farm mechanisation without realistically preparing the repayment schedule. Lack of supervision arising from the absence of technical staff also added to the problems of repayment irregularities. Commitment at the policy level to rural development through bank leading was hurriedly translated into action plans at the gross root levels with many aberrations and short comings. The seeds of loan default were sown, unwillingly. They germinated slowly, spreading all over. To assist the security-conscious bankers in expanding the agricultural advances the Credit Gaurantee Corporation of India Limited (1971) was established by Reserve Bank of India as a wholly-owned subsidiary institution, for a small premium the risk involved in financing the small farmers was covered partially.

Stage-2

The regional rural banks were establised during this period as another source of rural credit. By 1980, there were 85 which are operating in different parts of the country with 3279 branches. The District Credit Planning exercises initiated under the Lead Bank Scheme enchanced the involvement of the commercial banks considerably. The banker is now perceived as the provider of cheaper credit compared to the costlier non-institutional credit, the money-lender. Interest rates were regulated by Central Bank, linking them to the size of borrowings. The small borrowers could get credit as rates of interest between 10-12 per cent, while the prevailing rate in the rural sector was over 36 per cent. The banks now has the support of the field staff. The borrowers' perception has undergone a major change during this period. They have become conscious about the low cost of credit. They began availing loans for specific investments on land or on farm equipments. Their number increased rapidly. The banks, both the commercial banks and the regional rural banks–now have 0.86 crore borrowing accounts of farmers. The total advances rose to Rs. 2418 crores by 1980. Overdues became visibly an inevitable part of the rural credit structure. The amount of overdues was Rs. 447 crore, constituting 47.8 per cent of demand. Alarming signals have been beaming for quite some time.

Stage-3

A new phase in Indian banking started from October, 2nd 1980, with the implementation of IRDP. Through the quantum of credit involved in this programme is not very large, it made a perceptible change in the work schedule of the branch functionaries and the financial results of the branches. The subsidy oriented approach of the agencies administering this well-conceived programme started building up heavy pressures on the branch managers. The sub-targets for agricultural advances were received in April, 1980. The new directions were that the direct finance to agriculture (including allied activities) should reach a level of 16 per cent of the total advances by March, 1987; 17 per cent by March, 1989 and 18 per cent by March, 1990. The progressive upward revision of annual targets for agricultural advances as well as the ever-growing annual targets for IRDP borrowers delimited the horizon of the branch manager into the target given to him. He became the target of attack at the periodical review meetings, presided over by the district level officials for any short–fall in his achievements vis-a-vis the targets. His processes with targets inevitably leads to dilution in various processes of selection, documentation and supervision. Political pressure becomes inseparable for the populist poverty alleviation programmes. The borrowers' exposure to banks has tended to make them wiser. They are now seeking not cheaper credit but subsidy-linked credit. The overdues, Rs. 965 crores as on June 1996, were according for 38.1 per cent of the demand. Overdues under IRDP were much higher at 65.8 per cent for the year ending March, 1997. The poor recovery performance, if not adequatley improved, may have disastrous consequences when the coverage of families is widened on an on-going basis.

Recovery Strategies

Nearly one-fourth of the total gross NPAs were accounted by agricultural sector advances. In spite of the implementation of prudential norms and the gradual tightening of such norms, banks have been addressing the burden of NPAs through all the available remedies. As many as 17 of the 26 banks by the end of March, 98 (19 PSBs and 7 SBI Associates) had a net NPAs ratio of upto 10 per cent, while in the case of 9 banks, it was between 10 per cent to 20 per cent. However, the inadequate environment for enforcement of securities, the progress in recoveries has been slow. The difficulties and ineffective methods available for resolution of bad debts by banks are amply highlighted by the RBI studies. The data of 33 banks (27 public sector and 6 private banks) shows that by the end of March, 1998, over 46 per cent of the total quantum of NPAs were in suit-filed accounts. The recoveries made out of such suit-field accounts in March,

98 was as low as 4.3 per cent. The study found that over 14.37 crore suit-filled cases were pending disposal involving a shoping Rs. 21825 crore and added to that, a significant portion of such suits were pending for more than a decade. Hence, there is an urgent need to implement recovery strategies. The important links of the process are summarized below:

(i) The first step to be taken by branches for recovery is to organize a staff meeting wherein the recovery performance of the branch and the position of NPAs are discusses.

(ii) Immediately after the staff meeting, the branch managers in consultation with the credit officer/ manager credit forms a recovery team. The members of the team are to be chosen carefully and skills for performing recovery function and aptitude for field visits are to be given adequate weightage.

(iii) NPAs are classified into several categories by age, amount, location, financial health such as sick, compromise, recalled suit filed, decreed, execution of petition cases, etc.

(iv) Once the NPAs are properly classified and studied, each member of the team is given a certain target for recovery.

(v) After the target is set each member decides the recovery action to be followed. Recovery action starts with a reminder letter to be sent to a borrower. If there is no response, a second reminder is necessary.

(vi) All borrowers may respond to the reminder and therefore it becomes essential to visit them personally.

(vii) During the visit to the borrower's premise/ discussion with the borrower one may find that the borrower has genuine difficulties due to

which he is defaulting on payment of interest and loan instalments.

(viii) Whenever the branch staff does not convince the borrowers during the visits, it is attempted to bring pressure on them.

(ix) At any stage of the loan, compromise as a strategy of bank dues may also be thought of good borrowers who have defaulted on loan instalments due to genuine cases deserve certain concessions in interest dues.

(x) The process of recovery ends when the recovery of dues is fully realized.

Thus it can be concluded by quoting the Narasimham Committee 1998 which succinctly captures the import of resolution of NPAs: "NPAs constitute a real economic cost to the nation in that they reflect the application of scarce capital and credit funds to unproductive uses. The money locked up in NPAs are not available for productive use and to the extent that banks seek, to make provisions for NPAs or write them off, it is a charge on their profits. Banks have to charge their productive and deligent customers a higher rate of interest. It thus becomes a tax on efficiently. It is the customer who uses credit efficiency that subsidies the inefficiency represented by NPAs. This also raises the transaction costs in the system thus denying diligent credit customers the benefit of lower rates, which would help them to be more efficient and competitive. NPAs, in short, are not just a problem for banks, they are bad for the economy.

21

W.T.O.—A Boon or Bane to Agriculture

N. Satyanarayana
J. Vasantha Kumari

India's signing of GATT agreement in 1994 and joining of WTO as a founder member has put her agriculture into framework of global competition and the rule of global market. The central point of this development is competition which in turn depends on continuous improvement in efficiency. The basic theme of all the provisions of agriculture is to end the era of protectionist approach of development and to start a competitive, free and efficiency based growth. With the reduction in the level of tariff duties, customs and subsidies, the market access of agricultural products will increase and this, in turn, will benefit the agricultural sector. These provisions will make supply of agricultural products more competitive and will improve the efficiency.

Various studies (Gulati A. and others 1991, 1994, Nayyar, D. and Sen, A. 1994, Bhatia, M.S. 1994, Jakhar, B. 1993, 1994, Sharma, D. 1994, Gill, S. S, Brar, J.S. 1996, Kumar, T.K. and Ashok Mithal 1997, Ashima, G. 1997) which analysed the impacts of trade liberalisation policies on prices, costs, exports, marketable surpluses, subsidies, efficiencies of agricultural commodities did not yield encouraging results in the post-GATT period from the point of view of developing countries. Our products are comparatively costlier and our socio-economic conditions are not conducive to face competition of the world. Wheat, rice, tea are the most important exportable items and due to high cost and high price they are not internationally competitive. Besides, the issue of eliminating subsidies as well as imposition of intellectual property rights are also not favourable to Indian Agriculture. The worst sufferes are the two-thirds of small and marginal farmers.

Agricultural productivity in India is one of the lowest in the World and withdrawal of the subsidies would aggravate the situation, as the inputs would become costlier, increasing per unit cost of production even further. This would render the Indian foodgrains globally incompetitive. Free import of grains would also increase the price fluctuations in the economy and adversely affect the agricultural production and the food security. In Andhra Pradesh agriculture sector accounts for 1/3rd of domestic production, employs the bulk of the labour force and harbours majority of the poor.

Agriculture, according to the VISION 2020 is the "lifeline of Andhra pradesh's Economy", contributing "over a third of the states' GSDP and providing livelihood to over 70% of its population". Agriculture is considered as one of the growth engines and "by 2020 Andhra Pradesh will be power house of Indian agriculture, with a well developed efficient and productive allied sector and strong agro industries". "The agriculture will be shifted to a high growth trajectory". "...balanced regional development and equity will change the face of Andhra Pradesh". To achieve this vision, it is said, "the state will aggressively peruse strong agricultural growth", "with a quantum leap of 6% average growth rate in real terms over the next 25 years", and "A four-fold increase in agricultural GSDP".

The establishment of an agrarian economy which ensures food and nutrition to millions of people, raw material for its expanding industrial base and surplus for exports and fair reward system for the farming community, the services they provide to the society, will be main stay of reforms in the agricultural sector. In consonance with these objectives, the government of Andhra Pradesh has announced the new agricultural policy in the month of June 2000. On January 25, 2001, the Principal Secretary to government of Andhra Pradesh (Rural Development) had circulated the strategy paper on agriculture. The strategy paper on Agriculture lays emphasis on stepping up yields while cutting down production costs through a series of governmental and private initiatives to make farming a viable economic activity. The paper envisages a bigger role for contract farming, consolidation of lands at the farmer level and a major shift of labour from agriculture to services and industry. Keeping the crisis faced by the farmers due to the price crash in view, the strategy paper dealt with the issue of strenghtening the marketing network, creation of cold storage facilities, removal of movement restrictions and strengthening of Rytu bazaars.

What is Worrying the Farmers?

The declining prices of agriculture commodities in the demestic market is the major issue facing the government today. Adding fuel to the fire what is worrying farmers is that if they are unable to get good prices for their produce now, what will happen when the quantitative restrictions (QRs) on farm commodities are lifted under the agreement on agriculture (AOA) signed by India in the General Agreement on Tariffs and Trade? In past 50 years, despite exploiting all soft options, agricultural growth rate did not exceed 3.5 per cent. The scope for faster growth in future is very much limited. If the steps taken by the government increases the volume of production of agricultural commodities, when not backed by demand, can create 'gluts' in the market creating serious set back for further development. The "Vision 2020" presupposes automatic export of all products, assuming that markets are readily available. But this depends upon many factors which are not in the hands of state government. As it is, A.P. farmers are facing problems to market exportable products like tobacco, cotton, chillies etc. Exporters are getting lions share of profit in the times of plenty and at the time of falling prices, it is the actual producer, who have to bear the burden of the crisis. Even if India's exports boosted to 1 to 2 per cent share (from present 0.57 percent) A.P. exports will have to face stiff competition even from other states as it does not enjoy any premier proportion in terms of production and productivity in any of the commodities.

Agriculture sector, at present juncture, has been facing serious crisis, particularly in the production of rice, cotton, chillies etc. Hard hit are small and marginal farmers possessing less than 5 acres, whose holding constitute 81% in the total holdings under operation (1995-96 data). These holdings numbering 61.33 lakh (42.5 per cent of total land under cultivation). No attempt has been made to solve the problems being faced by them in order to make them economically viable and to boost production. No model will function with stability without making these holdings economically viable and sustainable. Relevant technologies and appropriate policy support of the state is needed to make these holdings economically self sustaining. When Beaurocrats Political leaders were not supposed to interfere in the decision-making process and allowing Co-operative organizations to operate on market based principles will go long way in improving the productivity of small farms. The forgotton agenda of land reforms must be revived and pursued vigorously not only for reasons of equity, but also for increased agricultural production. One of the most threatening impacts of WTO regime would be the easy availability of cheaper and better food grains. This would render the Indian farmers in a 'do or die' situation. Unless they are able to increase the productivity of their lands, they would not be able to compete with the cheaper imports with the present trend of increased prices of all inputs, withdrawal of fertilizer subsidy and higher irrigation charges (power charges also) the cost of production would skyrocket and it would be extremely difficult for the small farmers to survive in the market. Due to their backwardness in the field of agricultural research and development (R and D), most of the developing countries are bound to lose the battle of patents to the developed countries. It is likely that agriculture will survive in only those developing countries where the onslaught of free global economy that make large investments in the rural infrastructure, including creation of irrigation facilities, mechanisation of Agriculture, electricity, communications, modern technologies and new varieties with high yields. Hence our policies need to be reoriented and the focus has to shift from support price and subsidies to R and D, creation of rural infrastructure, increase in productivity and introduction of mechanical methods of farming in the backward areas. Supplementing govt. efforts, the private sector has to be encouraged to initiate R and D in the areas of biotechnology, seed development and tissue culture, so that high yielding varieties suitable to our conditions are developed. Nearly 60 per cent of the cultivated area in our state is dependent on rainfed farming. Without concerted action and development, there will not be any real development of agricultural sector. Therefore, rainfed farming based on watershed model should be encouraged. There are unlimited export possibilities for market of meat and their products, both domestic and export. The Government should make concerted effort to promote sheep and goat rearing. The policies like support price system and the distribution of the foodgrains through the PDS are myopic in nature and lack long term perspective, particularly under the scenario being offered by the WTO.

Hence, it is necessary that our policy-makers have to change their attitude towards anti-poverty and redistributive programmes in consonance with the requirements of WTO. The new farm policy must be forward looking, aiming at reduction of the miseries of the farmers, capable of attracting the required private investment and results in changing the cropping pattern as per the needs of the modern times. The threat arising out of WTO could hardly be visualised by our farmers. Now, when the situation is entirely injurious to our agriculture, it is entirely the responsibility of the government, to assist the helpless farmers get successfully out of the precarious situation of surplus foodgrains and impending WTO challenges.

REFERENCES

1. Vision—2020, Government of Andhra Pradesh.
2. Strategy Paper of Agriculture and Allied Departments, Government of Andhra Pradesh, January—2001.
3. Economic Survey—2001, Government of Andhra Pradesh.
4. Hindu Survey of Indian Agriculture—2000.
5. Gill S.S. and Brar J.S. (1996), Global Market and Competitiveness of Indian Agriculture, E.P.W., Aug. 10.

22

Economic Efficiency in A.P. Agriculture

P. Narasimha Rao
R.Venkata Rao

Economic efficiencies of the farm can be measured by comparing the actual profits with maximum possible profits on the basis of actual production function of the farm. The production function is to be estimated for the farms to identify the different variables which account for the differences in output and to assess the resource productivity between different categories of farms. Marginal value productivites of farms are to be computed to estimate the resource use efficiency at different levels of technology.

The estimaters of the parameters give the production elasticities of the factors included in the model. The elasticity of production of an input indicates the per cent increase/decrease in output for the increase or decrease in the quantity of that input for any specified level use of other inputs. If the coefficients of partial elasticities of production of all the inputs is less than unity with positive signs at all the levels of modernisation implies diminishing marginal productivity of factor inputs. In other words, by holding other inputs constant at their geometric mean level and increasing any of them, the yield would increase at a diminishing rate.

An attempt has been made in this paper to examine the resource productivity and resource use efficiency of the different categories of farms in different irrigated conditions through the estimation of production function. The data for this paper is drawn from an research project sponsored by U.G.C. The sample consists of 270 farm households and the distribution of sample households among different categoreis of farms in different irrigated conditions are presented in Table 1 sample households are divided into modern and traditional. The farms using modern techniques of new agricultural technology are termed as modern farms and the farms who do not use such type of technology termed as traditional farms. And the study area is divided into highly irrigated, medium irrigated and low irrigated based on the number of months the water is available. Both linear and non-linear forms of the production functions are tried. However, basing on the explanatory power of the model and significant levels of different parameters, the following Cobb-Douglas form of production function is selected:

$$\text{In Yi} = \text{In a} + b_1 \text{ In } x_1 + b_2 \text{ In } x_2 + b_3 \text{ in } x_3 + b_4 \text{ In } x_4 + b_5\ x_5 + e$$

Yi = Value of output (Rs.) per acre

X_1i = Value of seeds used (Rs.) per acre

X_2i = Value of fertilisers used (Rs.) per acre

X_3i = Value of pesticides used (Rs.) per acre

X_4i = Value of human labour used (Rs.) per acre

X_5i = Value of capital used (Rs.) pe acre

e = earror term

a = Technical efficiency co-efficient

$b_1b_2 \ldots b_5$ are production elasticity of the corresponidng inputs.

Table 1: Distribution of Selected Households by Size of Holding

Size Group	*Modern Farms*	*Traditional Farms*	*Total*
High Irrigated Zone			
Marginal	24 (44.44)	15 (41.67)	39 (34.33)
Small	17 (31.48)	10 (27.78)	27 (30.00)
Medium	10 (18.52)	8 (22.22)	18 (20.00)
Large	3 (5.56)	3 (8.33)	6 (6.67)
All	54 (100.00)	36 (100.00)	90 (100.00)
Medium Irrigated Zone			
Marginal	20 (37.04)	12 (33.33)	32 (35.55)
Small	14 (25.93)	10 (27.78)	24 (26.68)
Medium	15 (27.78)	11 (30.56)	26 (28.89)
Large	5 (9.26)	3 (8.33)	8 (8.89)
All	54 (100.00)	36 (100.00)	90 (100.00)
Low Irrigated Zone			
Marginal	16 (29.63)	10 (27.78)	26 (28.89)
Small	16 (29.63)	8 (22.22)	24 (26.67)
Medium	12 (22.22)	12 (33.33)	24 (26.67)
Large	10 (10.52)	6 (16.67)	16 (17.78)
All	54 (100.00)	36 (100.00)	90 (100.00)

The resource productivity for different categories of farms is examined with the help of above log linear Cobb-Douglas type of productive functions. The estimated values of elasticities for both categories of farms in three irrigated conditions are presented in Tables 2, 3 and 4. The estimated log linear Cobb-Douglas production function is found to be good fit as the variation in yield is explained by different factor inputs together varies between 65 per cent to 96 per cent in different irrigated conditions on both modern and traditional farms. Further most of the elasticity co-efficients have registered expacted signs and found to be significant at probability levels ranging from one to ten per cent. It may be observed from the tables that the inputs namely seed fertiliser, pesticides,human labour, and capital put together are explaining more than 50 to 90 per cent of the variation in output on both modern and traditional farms in three irrigated conditions. The intercepts of most of the estimated equations are positive except in medium irrigated zone. Further, most of the elasticity co-efficients have positive signs and are less than unity, indicating a decreasing marginal productivity of the factor inputs.

Table 2: OLS Estimates of Production Per Acre High Irrigated Condition

	Coefficient	*T-value*
1	**2**	**3**
Modernised Farms		
Constant	0.964	10.530
Seed	0.371*	8.545
Fertiliser	0.334**	2.34
Pesticides	0.034	0.535
Labour	0.315**	1.998
Capital	0.309*	7.849
Returns to Scale	1.363	-
R^2	0.96	-

(Contd...)

1	2	3
Traditional Farms		
Constant	1.454	2.65
Seed	0.238*	4.99
Fertiliser	0.298**	8.54
Pesticides	0.032	1.028
Labour	0.223**	2.004
Capital	0.088***	1.87
Returns to Scale	0.879	-
R^2	0.85	-

* Significant at 1 per cent level

** Significant at 5 per cent level

*** Significant at 10 per cent level

Table 3: OLS Estimates of Production Per Acre Medium Irrigated Condition

	Coefficient	*T-value*
1	2	3
Modernised Farms		
Constant	–1.724	6.38
Seed	0.123**	20.51
Fertiliser	0.456*	28.835
Pesticides	0.145***	1.86
Labour	0.245**	1.99
Capital	0.170**	2.354
Returns to Scale	1.139	–
R^2	0.90	–

(Contd...)

1	2	3
Traditional Farms		
Constant	–1.615	2.71
Seed	0.091**	1.69
Fertiliser	0.409*	51.52
Pesticides	0.116***	1.84
Labour	0.274*	2.99
Capital	0.108***	1.82
Returns to Scale	0.994	-
R^2	0.58	-

* Significant at 1 per cent level

** Significant at 5 per cent level

*** Significant at 10 per cent level

Table 4: OLS Estimates of Production Per Acre Low Irrigated Condition

	Coefficient	*T-value*
1	2	3
Modernised Farms		
Constant	2.525*	3.61
Seed	0.114	0.980
Fertiliser	0.545*	3.567
Pesticides	–0.138***	1.789
Labour	0.137**	2.36
Capital	0.045***	1.809
Returns to Scale	0.703	–
R^2	0.75	–

(Contd...)

1	2	3
Traditional Farms		
Constant	1.857*	1.482
Seed	0.125	1.135
Fertiliser	0.465*	15.502
Pesticides	–0.074	1.342
Labour	0.174**	2.289
Capital	0.023	0.447
Returns to Scale	0.712	-
R^2	0.65	-

* Significant at 1 per cent level

** Significant at 5 per cent level

*** Significant at 10 per cent level

It can be seen from the table that elasticity co-efficient of seed is statistically significant on both categories of farms of three irrigated conditions. The comparison of elasticity co-efficients of seed among different categories in three irrigated conditions shows elasticity co-efficients are higher on modern farms than on traditional except in zone III where elasticity co-efficient of seed on traditional farm is higher than the elasticity co-efficient of seed on modern farms. This implies that in medium and highly irrigated zones, the expenditure on seed exerts a relatively higher effect on income of modern farms as compared to traditional farms. On the other hand, the effect of the expenditure on seed on output is found to be not significant in case of modern farms of low irrigated zone and it is significant at 10 per cent level on traditional farms in low irrigated zone.

To sum up all the elasticity co-efficients of seed have positive signs and less than unity indicating a diminishing marginal productivity of this input. However, elasticity of output with respect to seed is higher on modern farms than on traditional farms with the exception of in low irrigated zone.

The results of the estiamted production functions reveal that fertiliser and FYM is the most important input to which output is more responsive in all irrigated situations and on both modern and traditional farms. The production elasticity co-efficietns of fertiliser and manures are statistially significant at 1 per cent and 5 per cent level of significance.The comparison of elasticity co-efficient indicated that elasticity co-efficients in modern farms are higher than traditional farms in all the irrigated situations. This reveals that output is highly responsive to the input fertiliser and manure in modern farms than on traditional farms in all the irrigated situations, i.e. increase in the expenditure on fertiliser and manure significantly increased the output more in modern farms than in traditional farms.

From the estimated production function it can also be observed that another important input to which output is highly responsive is expenditure on human labour. Human labour use in the production process is statistically significant in both modern and traditional farms of highly irrigated and medium irrigated situations. In low irrigated situation, the elasticity co-efficient associated with human labour is found to be negative and significant even at 10 per cent level in modern category and negative and not significant even at 10 per cent level in traditional category.

A comparison of elasticity co-efficients of human labour use among the different categories of farms for different irrigated situations showed that elasticity co-efficients are higher on modern farms than on traditional farms in high and medium irrigated situations. This indicate the fact that increase in the output is to be relatively higher by one unit increase in human labour use on modern farms than on traditional farms of highly irrigated and medium irrigated situations. In low irrigated situation, the elasticity co-efficient of human labour does not reveal any clear relationship with modernisation.

The elasticity co-efficient with respect to the expenditure on pesticides behaved differently under different irrigated conditions. The elasticity co-efficient of pesticide is positive and significant in modern and traditional farms of medium irrigated conditions and positive but not significant on both modern and traditional farms of high irrigated condition. The elasticity co-efficient of pesticide is negative on both modern and traditional farms of low irrigated situation and it is significant in modern farms and not significant in traditional farms at 10 per cent level.

A comparison of elasticity co-efficient of pesticides among different categories of farms under different irrigated situations reveal that output response to the pesticides is higher on modern farms as compared to traditional farms of highly irrigated and medium irrigated situations. This means that increase in output is higher in modern farms than on traditional farms for one unit increase in the expenditure on pesticides. The negative elasticity co-efficient in low irrigated situation indicate the fact that there is an excess expenditure on pesticides which is beyond the optimum requirement, so its contribution to output is negative. Thus modernisation has raised the output elasticity co-efficient of pesticide in highly irrigated and medium irrigated situations.

The elasticity co-efficient of capital is positive and significant among modern farms in all three irrigated situations. It is positive and significant in traditional farms of high and medium irrigated conditions, and positive and not significant in traditional farms of low irrigated situations.

A comparison of elasticity co-efficients of capital among different categories of farms under three irrigated conditions reveal that output elasticity of capital is high on modern farms than on traditional. This implies that the increase in output caused by one per cent increase in capital input for any specified level of use of other inputs, is higher on modern farms, than on traditional farms.

To sum up the production function analysis reveals that resource productivity is high on modern farms than on traditional farms. The comparison of elasticity co-efficients on both modern and traditional farms under three irrigated situations reveal that output, on an average, increase by larger proportions in response to a given percentage change in the expenditure on fertilisers and manures (with the inputs held constant) than to a similar change in other inputs. The elasticity co-efficietns with respect to human labour, seed, pesticides and capital assume same order of importance in the production process.

Returns to Scale

The returns to scale explain the behaviour of yield when all the inputs will change simultaneously in the same proportion. This is indicated by the sum of elasticity co-efficients of factors included in the Cobb-Douglas production function. Increasing returns, constant returns and decreasing returns to scalewill exists accordingly as the sum of elasticity co-efficients are greater than, equal to or less than one.

A comparison of the sum of elasticities reveals that the estimated returns to scales are higher on modern farms than on traditional farms in highly irrigated and medium irrigated situations. There is no significant difference on returns to scale between modern and traditional farms in low irrigated situation. A one per cent increase in all the resources will increase output by 1.36 per cent and 1.14 per cent in modern farms of high and medium irrigated situations, that is there are increasing returns to scale. There are decreasing returns to scale in traditional farms of high irrigated situation and approximately a constant returns to scale in traditional farms of medium irrigated situation. In low irrigated situation, there are decreasing returns to scale in both modern and traditional farms[1].

Resource Use Efficiency

The production function analysis is used determine the efficiency of resource use by estimating, the marginal value productivity of resources. Marginal value productivities are estimated at geometric mean level using the Cobb-Douglas type production function and is shown in Table 5, 6 and 7. The analysis reveals that marginal productivity of inputs are considerably higher on modern farms than on traditional farms. The marginal product for a given input indicates the expected increase in output forthcoming from the use of additional unit of that input, when the level of use of other inputs remains constant. The marginal productivity of any resource depends upon the quantity of the input that is being already used and the level of the other resources used with which are combined in the production process, so the estimates with widest acceptability are those derived at mean input levels with the stipulation that the most accurate estimates from the Cobb-Douglas production function are obtained with all input held at these geometric mean level.

It can be observed from the table that the marginal productivity of the input seed is 3.65 in modern farm and 2.45 in traditional farm of high irrigated conditions.While it is 4.60 in modern farms and 3.92 in traditional farms of medium irrigated condition, it is 4.66 in modern farms and 3.86 in traditional farms of low irrigated conditions. Further, the comparison of marginal productivities reveal that marginal productivity of the input seed on both modern and traditional farms is high in medium and low irrigated conditions when compared to high irrigated condition. This shows that a one rupee additional expenditure on seed adds more to the total value of output in low and medium irrigated conditions when compared to high irrigated conditions.

Table 5: Comparison of Marginal Value Productivity of Inputs Highly Irrigated Zone

	Moderns Farms		*Traditional Farms*	
	Marginal productivity inputs held at mean level of modern farms	*Marginal productivity inputs held at mean level of traditional farms*	*Marginal productivity inputs held at mean level of modern farms*	*Margtinal productivity inputs held at mean level of traditional farms*
Seed	3.65	4.33	2.10	2.45
Fertiliser	2.73	2.83	2.17	2.20
Pesticides	1.73	1.79	1.46	1.47
Labour	2.30	2.32	1.45	1.46
Capital	4.57	4.57	2.45	2.57

Table 6: Comparasion of Marginal Value Productivity of Inputs Midium Irrigated Zone

	Moderns Farms		*Traditional Farms*	
	Marginal productivity inputs held at mean level of modern farms	*Marginal productivity inputs held at mean level of traditional farms*	*Marginal productivity inputs held at mean level of modern farms*	*Margtinal productivity inputs held at mean level of traditional farms*
Seed	4.60	5.67	2.96	3.92
Fertiliser	3.18	3.68	2.47	3.09
Pesticides	2.41	2.15	1.66	1.56
Labour	1.30	1.30	1.25	1.31
Capital	2.25	2.34	1.23	1.28

Table 7: Comparasion of Marginal Value Productivity of Inputs Low Irrigated Zone

	Moderns Farms		*Traditional Farms*	
	Marginal productivity inputs held at mean level of modern farms	*Marginal productivity inputs held at mean level of traditional farms*	*Marginal productivity inputs held at mean level of modern farms*	*Marginal productivity inputs held at mean level of traditional farms*
Seed	4.66	4.08	3.41	3.86
Fertiliser	5.83	4.73	2.90	3.64
Pesticides	–0.97	–0.98	–0.39	–0.40
Labour	0.23	0.88	0.83	0.86
Capital	1.16	1.23	0.45	0.44

A comparison of marginal value productivities for modern and traditional farms in three irrigated conditions reveal that when the inputs are held at the geometric mean level of modern farms, a one rupee additional expenditure on seed increases the output by Rs. 3.65 in medium farms of highly irrigated situation and Rs. 4.60 in modern and Rs. 4.66 in low irrigated situation as against Rs. 2.10, Rs. 2.95 and Rs. 3.41 in traditional farms of high, medium and low irrigated conditions respectively. On the other hand, when inputs are held at the mean level of traditional farms, a one per cent increase in the expenditure on seed will add to income by 76 per cent more in highly irrigated condition, 44 per cent more in medium irrigated condition and 0.05 per cent more in low irrigated conditon.

Fertiliser

Marginal value productivitites of input fertiliser are higher in modern farms than on traditional farms in three irrigated conditions. A one rupee increase in the on fertiliser adds to the total value of output by Rs. 2.73, Rs. 3.18 and Rs. 5.83 in modern farms of high, medium and low irrigated conditions. In traditional farms one rupee increase in the expenditure on fertilizer adds Rs. 2.20, Rs. 3.09 and Rs. 3.64 to the total value of the output correspondingly in high, medium and low irrigated condtions. The marginal productivity of the input fertiliser on both modern and traditional farms is high in low irrigated condition, followed by medium irrigated condition and low in high irrigated condition. This shows that marginal productivity is increasing with the decrease in level of irrigation.

When the inputs are held at the generation mean level of modern farms a one rupee additional expenditure on fertiliser increases the output in modern farms by Rs. 2.53 in high irrigated, Rs. 3.18 in medium and Rs. 5.88 in low irrigated conditions. Where as in traditional farms at the same level of input use out-put increases by Rs. 2.17, Rs. 2.47 and Rs. 2.90 in high, medium and low irrigated conditions. Similarly when the inputs are held at geometric mean level of traditional farms a one per cent increase in the expenditure on fertiliser adds 28 per cent, 19 per cent and 29 per cent more to the income in modern farms of high, medium and low irrigated conditions respectively. When the inputs are held at the mean level of modern farms a one rupee increase in the expenditure on fertilisers is reduces the marginal productivity of fertiliser by 20 per cent, 22 per cent and 50 per cent in traditional farms of high, medium and low irrigated conditions respectively.

Pesticides

The marginal productivity of the input pesticide is higher on modern farms than on traditional farms except in low irrigated conditions. Here the marginal productivity of the input is negative on both modern and traditional farms of low irrigated conditions. This indicates that the expenditure on pesticides being used beyond optimum level in both modern and traditional farms of low irrigated conditions. The marginal productivity of pesticides on modern farms of high and medium irrigated conditions are 1.73 and 2.41 whereas it is 1.47 and 1.56 on traditional farms of high and medium irrigation conditions. A one rupee increase in the expenditure on pesticides adds to the total value of the output by Rs. 1.73 and 2.41 on modern farms of high and medium irrigated condition respectively. Whereas one increase in the expenditure on pesticides adds Rs. 1.47 and Rs. 1.56 to the total value of the output on traditional farms of high and medium irrigated conditions.

It can be also observed that when the inputs are held at geometric mean level of modern farms a one rupee increase in expenditure on pesticides increases the output on modern farms by Rs. 1.73, Rs. 2.41, in high and medium irrigated conditions respectively and at the same level of input use the increase in output on traditional farms are Rs. 1.46 and Rs. 1.66 in high and medium irrigation conditions respectively. Similarly when the inputs are held at geometric mean level of traditional a one per cent increase in the expenditure on pesticides adds 22 per cent and 38 per cent more to the value of output in modern farms of high and medium irrigated conditions respectively.

Human Labour

The comparison of the Tables 5, 6 and 7 reveal that marginal productivity of human labour is higher on modern farms than on traditional farms in high irrigated condition, whereas there is no significant difference in the marginal productivity of labour between modern and traditional of medium and low irrigated conditions. A one rupee additional expenditure on labour adds Rs. 2.30 in modern farms and 1.46 to income in traditional farms of high irrigated situation.

From the analysis, it may be noted that when the inputs are held at the geomatric mean level of modern farms a one rupee increase in expenditure on human labour, increases the output by Rs. 2.30, Rs. 1.30 and Re. 0.23 on modern farms of high, medium and low irrigated conditions. At the same level of input use, the output increases by Rs. 1.45, Rs. 1.25, and Rs. 0.83 on traditional farms of high, medium and low irrigated conditions. Further, the value of output increases by 59 per cent more in modern farms under high irrigated conditions corresponding to one per cent increase in the expenditure on human labour where the inputs are held at geometric mean level of traditional farms.

Capital

Further, it is observed that when the inputs are held at the geometric mean level of modern farms a one rupee increase in expenditure on capital increases the output by Rs. 4.57, Rs. 2.25 and Rs. 1.16 on modern farms of high, medium and low irrigated conditions. It increases the output by Rs. 2.45, Rs. 1.23 and Re. 0.45 in traditional farms of high, medium and low irrigated conditions at teh same level of input use.

A comparison of Tables 5, 6 and 7 reveal that marginal productivity of capital input is higher on modern farms when compared to traditional farms in three irrigated conditions. The marginal productivities of the inputs given in tables 5, 6 and 7 reveal that when the inputs are held at geometric mean level of modern farms, a one rupee increase in the expenditure on capital input adds to the value of output by Rs. 4.50, Rs. 2.25, and Rs. 1.16 in modern farms of high, medium and low irrigated conditions. It is Rs. 2.57, Rs. 1.28 and Re. 0.47 in traditional farms of three irrigated conditions respectively.

Further, when the inputs are held at geometric mean level of traditional farms a one per cent increase in capital input, on both categories of farms, the value of output increases by 84 per cent, 82 per cent and 179 per cent more in modern farms of three irrigated conditions respectively.

Thus the efficiency in the use of inputs seed, fertiliser, pesticides, labour and capital are high on modern farms than on traditional farms in all the three irrigated conditions. The elasticity of seed is highest in highly irrigated condition, and lowest in lowest in low irrigated condition. The efficiency in the use of fertiliser is high in low and medium irrigated conditions when compared to high irrigated condition. The elasticity of fertiliser is significant in all categories of farms in three irrigated conditions. The elasticity of pesticides is negative on both categories of low irrigated conditions and elasticity of pesticides is significant in both categories of medium irrigated condition. It is high in modern farms when compared to traditional farms. The efficiency in the use of human labour is high on modern farms than on traditional farms in all the three irrigated conditions and efficiency in the use of human labour varies with the irrigation levels.

The elasticity of capital is statistically significant in modern farms of high and medium irrigated conditions and traditional farm of medium irrigated conditions. The elasticity of capital input is high on modern farms and it is decreasing with decreasing levels of irrigation.

A comparison of return scale indicates that sum of elasticity co-efficients are greater than one in modern farms of high and medium irrigated conditions. This shows that there is increasing returns to scale in the modern farms of high and medium irrigated conditions. The sum of elasticity co-efficients is approximately equal to one in traditional farms of high and medium irrigated conditions indicateing constant returns to scale in these farms. And sum of elasticity coefficients on both categories of farms are less than one in the low irrigated condition indicating decreasing returns to scale.

Thus differences in marginal productivity of the same factor inputs between modern and traditional farms reflect the differences in Technology. These differences among other factors can be attributed to modernisation. This leads to the conclusion that for any given level of input use, the higher marginal productivity of inputs on modern farms gives higher output on modern farms when compared to the output on traditional farms at the same level of input use.

To sum up, although use of machines by itself may have no direct output increasing effect, but it plays a complementary or enabling roel in the production process which contributing towards increase in output along with irrigation. The modernisation enhances the capability to perform farm operations more intensively by filling the energy gap, when the stress is on economy of time as in multiple cropping.

Conclusions

1. The production function analysis reveals that resource productivity is higher on modern farms than on traditional farms. The elasticity of output with respect to seed, fertilizer and human labour is high on modern farms than on traditional farms.
2. The estimated return to scale reveals the existence of an increasing returns to scale in modern farms of high and medium irrigated zones. In traditional farms, a decreasing returns to scale exists in high irrigated zone and constant returns to scale is found in medium irrigation zone. In low irrigated zone decreasing returns to scale exists on both modern and traditional farms.
3. The analysis of marginal productivity reveals that marginal productivity of inputs is high on modern farms than on traditional farms. A one per cent increase in the expenditure on inputs added more to the value of output in modern farms than on traditional farms in three irrigated zones. The use of machines enhanced the ability of farms to perform farm operations more intensively by filling the energy gap, which cannot be met by animal or human labour when the stress is on economy of time in multiple cropping.
4. Modernisation of agriculture isa dynamic process, and its pace will be decided by the technical

upgradition of various farm technology and Govt. policies. Where there is a slump in recent years after green revolution. Thus a continuous review and development of technology must go along the way that satisfy the small and marginal farmers needs it should counter the needs of the small and marginal farmers in different agro climatic regions.

23

Agrarian Changes in Kaza, Anantavaram Villages of Guntur District: An Empirical Resurvey Study

A. Ranga Reddy

The history of all hitherto existing society is the history of class struggles.[1] The feudal production relations characterized by large land owners exploiting small farmers and landless labourers through narrowly–specified rigid contractual arrangements have often been considered to be the root cause of sluggishness in agricultural performance. The increased private and public investment in irrigation of recent years will certainly help to accelerate changes in agricultural production systems as indicated by the study villages. Real wages have increased over time and the labour market has tightened as employment opportunities both within and outside the village has expanded. As agriculture is further intensified in off farm and non-farm opportunities expand, the nature of land and labour markets is bound to be determined much more by market forces rather than the social and communal factors. Innovative technological policy and institutional interventions to exploit and encourage rural dynamism are needed to guide the process of economic growth.[2] The trouble with India is that there are too few exploiters and too many exploited siad Michal Kalecki.

The paper highlights the agrarian relations change in last twenty five years, especially the exploitation by large farmers over farm labourers.

Research Methodology

It is a matter of a great fascination that by re-visiting the same villages of Kaza and Anantavaram, which were surveyed by Marxist-Leninist analyst, Puchalapalle Sundarayya in the first quarter of 1974 year. After a quarter century, we visited these villages during the period of July 1997 to June 1999 and canvassed a questionnaire on August 1999 using same methodology with certain modifications in concepts like Farm labourer (0.00), Marginal farmer (0.00-1.0 acre), Small farmer (1.0-2.0 acres), Medium farmer (2.0-4.0 acres), Large farmer (4.0-10.0 acres) and Big farmer (10.0 and above acres) which are popularly used in Government and Research Institutes in India.

To examine the application of Marxian doctrine to agriculture, he had his team had chosen Kaza as dry and Anantavaram as wet farm management villages in Guntur District of Andhra Pradesh and taken the conversion of 2.50 acres dry as equal to 1.00 acre of wet land on assessing derivation of income.

At present, we observed that in Kaza dry land is under heavy demand and shortage, whereas its current market price is found more than wet land, as it is usually availed for growing commercial crops. So that we considered all lands of both villages as wet. In order to focus on agrarian changes, we disregarded the non-farm categories, absent of heads of households as on that survey day and adopted **census** method.

To cross check these results, we took facts and figures from the village Administrative Officer, Executive Officer– of these village Revenue and Panchayats' records respectively. Besides it, in deductive approach, we interviewed elderly, experienced farmers to get their impressions views over the past and present transformations in farming activities to strenghten the reliable data extracted directly from farm households.

Objectives: The followings are the major objectives of this study:

1. To identify the economic class differentiation among the peasantry of villages.
2. To find out the extent of land concentration among dominant castes.
3. To examine the pattern of tenancy among farming community.
4. To assess the exploitation of wage labour by the bigger farmers.

Kaza

It is situated 15 kms distance to Guntur on 5th Highway route to Vijayawada. Growing population size in the village is ever raising due to University, Factory and Trade activities. Increasing population at the rate of more than two per cent per year is a regular phenomena on par with the country and also A.P. State. The Table 1 shows the composition of population in Kaza village.

Table 1 provides an unique picture of 69 per cent of O.Cs, followed by 16 per cent of B.Cs and 11 per cent of S.Cs. A village or a household has been trying to exploit the land potentiality in optimum level, so that the village or the concerned household will get better income. This village has got number of advantages in obtaining windfall profits. Owing to placed in Highway and in between two big cities, land got attractive appreciation in its value. The table highlights the land use pattern.

Table 1: Kaza Population 1991

Category	*Population No.*
ST	250 (4.0)
SC	648 (11.0)
BC	950 (16.0)
OC	4,108 (69.0)
Total	**5,956 (100.0)**

Source: Population census 1991.

The Tabel 2 emphasizes that 70 per cent of land is utilized for sowing purpose and out of it, 40 per cent is for double cropped area. During the last 25 years, entry of Guntur Channel and Krishna Canal water into the fields of this village had changed the face of the village in agriculture, in particular and overall economic scenario, in general. Slowly, irrigated land is going for Housing plots.

Table 2: Land Use Pattern in Kaza-1998-99

(in acres)

Barren Land	503 (11.0)
Forest	-
Plots	392 (9.0)
Sown area	3,133 (70.0)
Quarry	467 (10.0)
Total	4,495 (100.0)
Double Cropped Area	1793 (40.0)

Source: VAO: Revenue records of Kaza

The crop pattern is an indication of development paradigm, which is presented.

Table 3 shows that 76 per cent of irrigated area utilized for paddy, blackgram and vegetables and the rest was used for commercial crops. Irrigated dry and private motor pump-set owners favoured commercial crops of vegetables because of higher prices and instant income for the products in the market. Vegetable traders are touching this village market everyday and paying spot high prices. Irrigation is a perennial boon, which is shown in Table 4.

Table 2: Crop Pattern in Kaza 1998-99

(in acres)

Paddy	1583 (50.0)
Cotton	600 (19.0)
Mirchi	350 (11.0)
Blackgram	150 (5.0)
Vegetables	450 (15.0)
Total	**3,133 (100.0)**

N.B.: Figures in the parentheses shows the percentage to the total

Source: VAO—Revenue Records of Kaza

Table 4: Irrigated Area in Kaza

(in acres)

Guntur Channel	*Wet*	*Irrigated Dry*
Lift Irrigation	240	450
Krishna Canal	852	-
Total	**1092**	**450**

Source: VAO: Revenue Records of Kaza

Anantavaram

This is situated 13 Kms distance to the east of Tenali Town. Land is the prestigious and productive asset in the village. Due to the totally irrigated area, the prosperity was concentrated among upper caste (Sudra) households only.

Table 5 shows very clearly that the village lands are absolutely irrigated. Due to long period of irrigation, around 20 per cent of soils are being found in salinity and water-logging. Growing Paddy as a staple crop is popular since more than a century. Recently, when state supported a sugar factory near Jampani, is founded, slowly the large farmers switched over to grow sugarcane crop, for a change.

Table 5: Land Use in Anantavaram 1998-99

(in acres)

Barren	Nil
Forest	-
Irrigated area	2,450
Residential area	50
Geographical area	2,500

Source: VAO—Revenue Records, Anantavaram

For milching animals, green grass feed is necessary. Therefore, few small and large farmers have allocated some piece of land for Junnu grass.The excess grass will be sold in the market in the form of bundles. The crop pattern is given below:

Table 6: Crop Pattern in Anantavaram 1998-99

(in acres)

Paddy	1,950 (80.0)
Sugarcane	400 (16.0)
Grass	100 (4.0)
Total	**2,450 (100.0)**

Source: VAO—Revenue Records, Anantavaram

Table 6 reveals that 80 per cent of the land is being used for double cropped paddy. 16 per cent of land was availed for sugarcane growing since 1997. Guaranteed income is enjoyed by this villagers since-one and half century. By examining the infrastructure facilities in the village, it had all minimum facilities. It had–protected drinking water, health centre, telecommunication,transport, library–facilities to make people more efficient and productive. This village is called as urbanised rural village.

Irrigated Revenue: This village is paying revenue of Rs. 4,25,000/- to the State exchequer for using the irrigated water every year. The payments are prompt and perfect. The village had a permanent revenue collection office with required staff.

Caste: India is a caste-ridden society. Every region had their own caste configuration. Some castes are in dominance and other are in suppressed. During the last fifty years, some more caste households emerged as economically stronger, due to irrigation, contracts, state employment, subsides etc., upper castes i.e. Brahmins, Kammas were possessed more land and rest of the castes had meager land. However, since seventies Brahmins, sold away their land and moved to urban areas for their settlement. Due to this, land has been transferred to the backward caste people since they have purchased.

Farm Labour Market: Labour is considered as a key input for smooth functioning of agricultural operations. The elasticity of wage rate in the labour market would be fixed by demander. In classical theory, as the wage rate increases more number of labourers would show willingness to work. In the current period, a new system is emerged, i.e., contract system. In this system, labourer will work few hours and get more wage rate than casual/regular labourer. The community/farmer has to give the piece of work in wholesale manner by fixing a price. Many of the operations, which are essential, are found in the list of contract. The unfortunate thing is that, even in small farmers' category also family labour's contribution is declining. Crop cutting machines are entering in this village, inspite of resistance by labour community through **communist party**. In general, anti-proletarianization trend is found in the village that means, shifting labour class to marginal or tenant class, that is, securing a piece of land for income[6].

Land, the most scarce factor of productions is considered as the only important resource in agrarian society, not only for livelihood bur for social recognition and status also. Land distribution is highly skewed in favour of large and big farmers. The iniquitous distribution of land implies varying economic positions of the participants in the lease market. Fragmentation of holdings is one of the factors that reduces the efficiency of agricultural productions and consequently the landowners prefers to lease out the fragments to those with adjacent holdings. The changes in caste-class reflect the class-reproduction strategies of employers-employers, both internally differentiated, but contracts are naturally heavily influenced by the requirements of the dominant class (Ravi Srivastava, IJAE, 2000).

Class Differentiation: In earlier days the villages essentially consisted of a cultivation–based nucleus involving Brahmins, Vyshyas, Sudras, combined with a collection of relatively small service castes linked to that nucleus through the **Jajmani system.** This provides the background to the profound changes that have taken place over the period.

Providing irrigation, intervention of economic programmes of state like IRDP, JRY, DWACRA etc., Financial institutions penetration, links with urban economy have been served as a catalyst to change the class and land transactions among all castes. The Table 7 explain the current changes.

Table 7 denotes that in Kaza two acres (SF) landed holders are 33 per cent among dominant communities, where it was, earlier in 1974 study 53 per cent. Among

Table 7: Economic Differentition Between Different Communities

	Kaza						
	FL 0.0–1.0	*MF 0.1-0*	*SF 1.0-2.0*	*Med.F 2-4*	*LF 4.0–10.0*	*BF 10.0 & above*	*Total*
OC	49 (19.0)	72 (28.0)	53 (20.0)	55 (22.0)	25 (10.0)	2 (1.0)	256 (100.00) (46.00)
BC	66 (26.0)	114 (44.0)	45 (17.0)	17 (7.0)	15 (6.0)	1 (0.0)	258 (100.00) (47.00)
SC	1 (7.0)	11 (79.0)	0.0	2 (14.0)	0.0	0.0	14 (100.0) (3.0)
ST	19 (86.0)	3 (14.0)	0.0	0.0	0.0	0.0	22 (100.0) (4.0)
Total	**135 (25.0)**	**200 (36.0)**	**98 (18.0)**	**74 (13.0)**	**40 (7.0)**	**3 (1.0)**	**550 (100.0) (100.0)**

(Contd...)

Table 7: (Contd...)

	Anantavaram						
	FL 0.0	*MF 0.0-1-0*	*SF 1.0-2.0*	*Med.F 2.0-4.0*	*LF 4.0-10.0*	*BF 10.0 & above*	*Total*
OC	22 (15.0)	25 (8.0)	28 (20.0)	25 (19.0)	31 (22.0)	10 (7.0)	141 (100.00) (36.00)
BC	48 (54.0)	24 (27.0)	6 (7.0)	5 (6.0)	6 (6.0)	.0	89 (100.00) (23.00)
SC	54 (72.0)	18 (25.0)	1 (1.0)	0.0	1 (1.0)	1 (1.0)	75.0 (100.0) (19.0)
ST	84 (98.0)	2 (2.0)	0.0	0.0	0.0	0.0	86 (100.0) (22.0)
Total	**208 (53.0)**	**69 (18.0)**	**35 (9.0)**	**30 (8.0)**	**38 (10.0)**	**11 (2.0)**	**391 (100.0) (100.0)**

Source: Field data.

classes of upper caste (O.C.), Marginal Small farmers are found in 48 per cent, however they are 61 per cent in backward class.

In Anantavaram, then above 2.5 acres among all categories 40 per cent of upper caste, 6 per cent BCs and no SCs were recorded. Present survey have shown that 48 per cent of upper caste, 12 per cent of BCs and 2 per cent of SCs.

Regarding Sundarayya study of 1974, proved that in Kaza, above 2.5 acres category 53 per cent of dominant caste–landlord and arich peasant class suppresses 60 per cent of all the families belonging to backward communities SCs and STs for success of working class movement it is better to be enlisted the support of the 14 per cent well to do of BCs and forge an unity with poor and middle peasants, to win over a neutralize the better off sections in the dominant section. In present economically developing Kaza 47 per cent BCs are emerged as stronger said group than OCs. Among 43 large and big farmers, 37 per cent belong to BCs alone. As OCs are the migrants to the nearest urban areas, BCs are found in farming class and ruling class, as a Sarpanch of the Panchayat.

In earlier study of Sundarayya 1974, proved that in developed village-Anantavaram 68.6 per cent backward household communities and 95 per cent of SCs were found in landless category and owners of 0.5 acre and less, whereas the category of 10 acres and above is predominantly dominant communities. Whereas today the survey discovered that 36 per cent OCs (upper class) still marginally dominates in over all social communities. Nonethelss, 23 per cent of BCs would overshadow in nearest future. Tough competitive economic life is a hallmark of this village among SCs and STs also. Gradually the ownership of land is passing from dominant castes to BCs and SCs significantly 6 large and one big farmer in BCs and one big farmer in SCs.

Land Concentration: Land ownership changes at the household level in two distinct transfers-land sales and purchases and the process inheritance and partition. Even in longer periods, land sales and purchases have had a secondary influence in comparison with the process of inheritance and partition. The land endowment of few households had depended for more on the division of land with in such a dynasty than on transaction between dynasties. The Table 8 finds the latest position.

Table 8 reveals that the average per household land holdings differs, such as 1.61 acres in developing village and 1.75 acres in developed village. Marxian analysts felt that concentration of land in farming leads to exploitation of labour either in working hours or wage rate or other socio-political areas. Then study found on Kaza, that above 5 acres of land, 5.2 per cent of families possessed 34 per cent of land and landless or less than 0.5 acres, 44 per cent of families owned 4.4 per cent of land, which was an extreme inequality. As on today, 61 per cent of households are below 1.0 acre, above 4 acres, hold 18.0 per cent of land and 8 per cent of households possessed 33 per cent of land. Nonetheless, land is started to percolated among downtrodden.

Then in Anantavaram, above 10 acres, constituting 5 per cent of farmers owned 53.2 per cent of land and 50.3 per cent of households carried no land at all. Land is concentrated among bigger landowners especially dominant communities. At present, above 10 acres, consisting of 3 per cent of farmers controlled 30 per cent of land and 53 per cent landless labourers have no land. In this village inequality is raised more than earlier study.

Tenancy: Under feudal mode of productions, the terms and conditins are exploitative and coercive against tenants (Bell and Srinivasan 1989 Chadha and Bhaumik 1992; Mahapatra 1994). Tenancy spins a major roel in these two

Table 8: Land Concentration

	Kaza			
Households and Land / *Category (in acres)*	*No. of Households*	*Percentage to total households (in acres)*	*Owned land by these household (in acres)*	*Percentage to total owned land*
Farm Labour (0.0-0.00)	135	25.0	0.00	0.0
Marginal Farm (0.0-1.0)	200	36.0	157.75	18.0
Small Farmer (1.0-2.0)	98	18.0	179.56	20.0
Medium Farmer (2.0-4.0)	74	13.0	254.65	29.0
Large Farmer (4.0-10.0)	40	7.0	253.40	28.0
Big Farmers (10.0 and above)	3	1.0	43.00	5.0
Total	**500**	**100.0**	**888.36**	**100.0**

(Contd...)

Table 8: (Contd...)

	Anantavaram			
Households and Land / *Category (in acres)*	*No. of House-holds*	*Percentage to total house-holds (in acres)*	*Owned land by these household (in acres)*	*Percentage to total owned land*
Farm Labour (0.0-0.00)	208	53.0	0.00	0.00
Marginal Farm (0.0-1.0)	69	18.0	53.11	8.0
Small Farmer (1.0-2.0)	35	9.0	67.40	10.0
Medium Farmer (2.0-4.0)	30	7.0	94.10	14.0
Large Farmer (4.0-10.0)	38	10.0	265.00	38.0
Big Farmers (10.0 and above)	11	3.0	207.60	30.0
Total	**391**	**100.0**	**687.21**	**100.0**

Source: Field data.

village agricultural economies, mainly as a convenient means for bringing together the relevant factors of production. Leasing was common at all levels of land ownership scale, with overall difference in land ownership between tenants and landlords/large farmers. Among two types to tenancy-cash/kind rent and share cropping contracts, the former become a popular. Upper caste households are partially or fully played the role of rent-seekers and lower cast households as rent givers. The Table 9 provides a birds eye view.

Table 9 spells out that in developing village per household leased in area is 2 acres, whereas in developed village it is 3.25 acres among dominant classes. The fact is that transacting land in between landlord and tenant among upper class and SCs. If upper classes believable tenant is not available in upper class, they prefer BCs next to SCs. This is the order of the day. Marxist treatment of rent give considerable attention to the evolution forms of rent. The emergence of fixed cash rents, which transfer production risk to the tenant but allow the tenant to treat the surplus, net of the final sent as profit, is considered to be more compatible with the developemtn of capitalism in agriculture.

In earlier period, in Kaza 28 families holding in between 10-15 acres each were treated landlords leased out an area of 73.25 acres. At present, 27 families holding 70 acres of land leased out. For rent, each acre may get cash rent of around Rs. 5,000 or kind of 16 to 18 bags of paddy. 8 per cent of land, out of 886.36 acres was leased out and 9 per cent was leased in by upper class category. Out of 258 B.Cs households, 10.4 per cent households leased out 35.50 acres of land, in it 25 per cent of households leased in 122 acres of land. In overall leased in land of 206.5 acres, B.Cs alone took lion-share of 59 per cent.

Regarding Anantavaram area, then 17.3 per cent or 288 acres of total cultivated area, are under lease on rental basis. The number of families which lease in land was 149,

Table 9: Caste-Wise Tenancy: Leased Out and Leased In

			Kaza			
Tenancy Caste wise	*Total house-hold (No.)*	*No. of House-hold leased out*	*Leased out area (acres)*	*No.of House-hold leased in*	*Leased in area (areas)*	*Average per house-hold leased in area*
OC	256	27 (48.0)	69.50 (65.0)	41 (37.5)	80 (38.75)	2.0
BC	258	27 (48.0)	35.50 (33.0)	65 (60.0)	122 (59.2)	2.0
SC	14	2 (4.0)	1.50 (2.0)	2 (1.5)	4 (2.0)	2.0
ST	22	0	0	1 (1.0)	0.50 (0.25)	0.50
Total	**550**	**56 (100.0)**	**106.50 (100.0)**	**109 (100.0)**	**206.50 (100.00)**	-

(Contd...)

(Table 9: Contd...)

Tenancy Caste wise	*Total house-hold (No.)*	*Anantavaram*				
		No. of house-holds leased out	*Leased out area (acres)*	*No. of House-holds leased in*	*Leased in area (acres)*	*Average per household leased area (acres)*
OC	141	32 (60.0)	191 (80.0)	16 (26.0)	52.0 (37.0)	3.25
BC	89	15 (28.0)	41 (18.0)	24 (39.0)	56.0 (40.0)	2.33
SC	75	5 (10.0)	5 (1.60)	17 (27.0)	27.5 (20.0)	1.6
ST	86	1 (2.0)	1 (0.40)	5 (8.0)	5.0 (3.0)	1.0
Total	**391**	**53 (100.0)**	**238 (100.0)**	**62 (100.00)**	**140.5 (100.0)**	**-**

Source: Field data.

constituting 18.2 per cent of all the families in the village. As on today, out of 141 households in upper classes, 23 per cent of households are in leased out and 80 per cent of the total land of 238 acres is leased out, whereas 11 per cent of households of the total area in leased in showing 37.0 per cent of the leased in area. Among 89 B.Cs households 17 per cent of households are in leased out for rent purpose, the area of land is 18 per cent of leased out, while 27 per cent households took leased in land of 40 per cent of total leased area. In both the villages B.Cs emerged as the strong group in taking land for lease in, whereas upper class was found in leased out category. SCs also are found in third grade in lease in category in this developed village.

Distribution of Farm Labourers: Perhaps, modern farming and post-economic reforms have taken a strong impetus to push more farmers of low land holdings or landless labourers into poverty and unemployment, which are in particular class and perennial in Coastal Andhra districts. The Table 10 formulates the issues.

Table 10 reveals that child labour is existing in developing village which is on decline in developed village. Pure wage earners are found in B.Cs of 65 per cent in developing village and 63 per cent among SCs of developed village.

Former study in Kaza proved that 149 daily rated agricultural families of whom 62 are landless, 67 own less than 0.5 acre and 20 between 0.5 and 1.0 acre. These 149 families account for 14.5 per cent of the total, own 36.6 acres (2.3 per cent) and get employment for 253 days per family. They make 5.3 per cent of the total village income, with an annual average of Rs. 730 per family, that is, Rs. 204 per capita or 56 paise per capita per day. At present, out of 550 examined households, 237 farm households are engaged in farm labour income, that is 43 per cent are associated with wage labour. 60 per cent of Backwared class households, 56 per cent of total farm labour land was used

Table 10: Distribution of Farm Labourers

Wage earness / Caste wise	Kaza						
	No. of households	Among wage earner households	Land	Total househld members	Wage earner Households		
					Male	Female	Children
OC	256	57 (24.0)	84 (40.00)	284 (26.0)	78 (2.0)	70 (28.0)	11 (31.0)
BC	258	154 (65.0)	118 (57.0)	693 (65.0)	164 (62.0)	155 (63.0)	23 (66.0)
SC	14	23 (10.0)	6 (2.50)	90 (8.0)	22 (8.0)	20 (8.0)	1 (3.0)
ST	22	3 (1.0)	1 (0.50)	15 (1.0)	3 (1.0)	2 (1.0)	0
Total	**550**	**237 (100.0)**	**209 (100.0)**	**1082 (100.0)**	**267 (100.0)**	**247 (100.0)**	**35 (100.0)**

(Contd...)

(Table 10: Contd...)

			Anantavaram				
Waage earness / *Caste wise*	*No. of house-holds*	*Among wage earner house-holds*	*Land*	*Total house-hold members*	*Wage earner Households*		
					Male	*Female*	*Children*
OC	141	15 (10.0)	8 (30.0)	66 (12.0)	15 (9.0)	11 (7.0)	0
BC	89	37 (26.0)	19 (70.0)	148 (26.0)	38 (22.0)	31 (21.0)	3 (43.0)
SC	75	63 (43.0)	0	178 (32.0)	72 (43.0)	64 (44.0)	4 (57.0)
ST	86	30 (21.0)	0	168 (30.0)	45 (26.0)	42 (28.0)	0
Total	**391**	**145 (100.0)**	**27 (100.0)**	**560 (100.0)**	**170 (100.0)**	**148 (100.0)**	**7 (100.0)**

Source: Field data.

to get an average income of Rs. 9,000 per annum for male and Rs. 5,400 per annum for female. It is found in general that male labours was paid Rs. 50 per day, female labourer Rs. 30 and for a boy Rs. 30 to 50 range for all farm operations, Maximum labour days per person were calculated around 180 per annum and 90 minimum. Job holders in the University, tobacco processing unit, barytes cutting, hotels, market yards, panchayat, party offices are excluded in this study.

Earlier study on Anantavaram had investigated that average annual income per family for the 195 daily wage earners is Rs. 657 and the per capita income a meagre of Rs. 200. Only those 55 families which possess some tiny plot of land of 0.25 to 0.50 acre or those which have leased in an acre or so, get incomes varying from about Rs. 860 to Rs. 1200. At present, out of 391 examined households, 145 households, taht is, 38 per cent households are associated with farm labour to get wage rate, apart their piece of productive land. SCs are found more in wage earning group, follows BCs and STs. In this developed village also wage rates per day is Rs. 50 for male, Rs. 30 for female and Rs. 30-50 to a boy. Active labourer would get 180 days per annum and 90 days minimum. Average income of the farm labourer for a period of 180 days is estimated on Rs. 9,000 as in developing village which is below poverty line. Workers in Panchayat, tanks, parties, teachers, employees, retired are excluded. Working days have fallen steeply, because of introduction of machinery in many of farm operations like transporting, spraying, rice planter, harvesters, but raised a wage rate, which is marginal, comparing market wage goods inflation.

Conclusion

It is found very clearly that ownership of land is passing from dominant castes to Backward and Scheduled Castes. Among the villages the land is not trickle down to the lowest stratum, but inequality is widened. Though much of area was sowed in Kaza, 70 per cent of households are found in farming with high degree of social backwardness, whereas in Anantavaram many of the farm owners turned into landlords and engaged intrade at Towns. In both villages education was not penetrated very much except upper caste households of Anantavaram. The workers in agriculture are inclined to shift from agriculture to non-agriculture. In developing village, the surpluses are invested on housing, whereas in developed village, they invested more on land alone, which is more productive. In both the villages BCs were emerged as the strong group in taking land for lease in, whereas upper class was found in leased out category. The negative attitude of exploitation was on diminishing among upper classes over working group. In earlier days, Annual Farm servant system was observed for 365 days, whereas to day it come down to 180 days per year.

REFERENCES

1. Karl Marx and Friedrich Engels: The Communist Manifesto, Washington Square Press, New York, 1964, p. 57.
2. Vishwa Ballabh and Sushil Pandey: Transitions in Rice Production Systems in Eastern India. Evidence from the villages in U.P., *Economic and Political Weekly,* March 27, 1999 p. A. 15.
3. Jean Dreze, Peter Lanjouw and Naresh Sharama: Palanpur, 1957-93 Part–II: Economic Development and Social Change, Centre for Development Economics, Op. No. 4, p. 26.
4. K.P. Kannan: Levelling up or lavelling down? Labou Institution and Economic Development in India, *Economic and Political* Weekly, 23 July 1994.
5. Nicholas Stern and Peter Lanjouw: Economic Development in Palanpur over Five decades (ed) Clarenden Press, Oxford 1998.

24

WTO and the Indian Agriculture

L.S.N. Prasad
P. Vasudeva Rao

There is a lot of hue and cry against WTO (World Trade Organization) in India. Even after accepting the international trade norms and joining in WTO some politicians, some of the intellectuals and the media have been focusing the negative aspects of two rules and regulations. The criticism that much has been levelling against WTO is only in our country. There are number of countries which opposed the WTO norms includes USA, EEC, but they successfully achieved their nations objectives to a large extent.

The WTO is a legal authority at international level equipped with enforce met rule (not mechanism till today) to maintain and decline the international tradition system. After prolonged GATT negotiations the world body was established on January 1, 1995. According to two constitutions and GATT rules the below articles are bonded to the member countries in international trade. They are:

Article XI: General elimination of quantitative restrictions on member states.

Article XII:	Restriction to safeguard the balance of payments of the member countries
Article XVI:	Subsides granted by the nations to other countries
Article XIX:	Emergency action on imports of particular product
Article XXIV:	Territorial application of frontier traffic and customs and free trade areas
Article XXVII:	Modification schedule of tariff reduction
Article XXXVI:	Trade and development

With the above article that are inscribed in the constitution curbs the exploitation of any member country by the other country. Unlike the previous negotiation agriculture was introduced tint he international trading negotiations in Ts because of the huge subsidies given by the EU.

There is erroneous view that in our country the Uruguay Round agreement has forced our country to cut back the subsidies on agriculture. By this agreement we are also forced to shift away from food grains in large scale which leads to no market for domestic production.

The Reality in Agricultural Subsidies

Due to agreement on agriculture (AOA), India could not remain more aloof from the rest of world so it has decided to join in WTO. This AOA has, established a number of generally applicable rules with regard to agriculture and trade related matters. Most of the developed countries are providing huge subsidies to the agricultural sector. In these countries a very low than per cent of the population are dependent on agriculture. The following table gives the information about the subsidies that are being given to agricultural by different countries and the per cent of population depending on agriculture:

Country	*Subsidy per hectare*	*% subsidies*	*Population depending on agriculture*
EEC	$82	37%	8%
USA	$32	26%	5%
Japan	$35	72%	4%
China	$30	34%	24%
South Africa	$24	60.67%	18%
India	$14	2.33%	60%

Source: WTO report.

From the above table India provides only 2.33 per cent subsidies to agriculture. It is very less when we compare with other countries. The agreement on agriculture will allow the Aggregate Measure of Support (AMS) to agriculture will be 10 per cent in India and it would be 5% in developed world. The total subsidy on agriculture in the developed works is $ 150 billion and in developing world I $ 19 billion. If the AMS exceeds 10 per cent in any country, it has to reduce by 13 per cent by 2004. According to WTO the AMS in India, the non–product subsidies is 7.5 per cent and subsidy on non product is –38.5 per cent. So there is no question of losing anything in agriculture trade by accepting AOA.

Tarrifs: There was a lot of change in the restructures of tariff on agricultural commodities in compliance of WTO agreement. India commits to reduce tariffs on 686 agricultural products. The average tariff on agricultural commodities are 115 per cent before the agreement. After the agreement it is reduced to 35 per cent. Along 686 commodities the tariff on 587 commodities is on an average is 50 per cent less than the negotiate tariff tee. Only on 10 commodities the tariff rate is more than the negotiated tariff rate.

We have to recall that in this contest 2 decades ago India agreed at intentional meeting that it will reduce the tariff either to '0' are to minimum level on the important of dairy products. So it is not now that we agreed the import of agricultural products. In this context of edible oils we agreed (binding) to impose tariff up to 150 per cent but the bounded duty at present is '0'. With the duty reductions of tariff on edible oils as low as 25 per cent. By this hundreds of oil mills were closed down, ground nut farmers are unable to get minimum price. Some times they disposed their product at huge losses. The pulses were imported under OGL at zero interest rate. In July 1999 the government of India renegotiated to impose the tariff on pulses. But the international community has not yielded.

The Challenge

(a) ***Investment:*** Indian agriculture needs a lot of investment to have large scale production. By this we can reduce the production cost. But the problem lies with the employment to the displaces persons who are depended on agriculture. Instead of criticism the WTO (.) we have to think about the providing of alternative system. Corporatisation of agriculture is against the social justice. We are in the process of distribution of land to land less poor by implementing the land ceiling acts. It will be a hurdle to improve the production as well as productivity.

(b) ***Patient system:*** The plant such as neem, turmeric and products such as Basmathi (Scented rice) are patented or about to to American companies. It is a peculiar that in India atheism plants ace being used in the entire households in the country since ancient times. Giving a latent to the America based company is, where the country's existence is 3 centuries back, not accepted and foe theta we ace

to fight. There is dispute settlement board (DSB) at Geneva. The Indian Law people are not aware of the international law and Laos it is costlier.

Allowing the Foreigners to Fish in Indian Waters

There is another problem that was arise after accepting the WTO norms. Every country should allow the other country to catch the fish in its territorial waters. India is one of the exporters of special grade fish (TUNA) and prawns to other countries and earns substantial amount of foreign exchange. Around 10 lakh people are depending on the activity of catching fish. If we allow the foreigners, they catch fish in deep sea waters (deep sea fishing) which leads exhausting of fish resources, but also displacement of the people in fish catching, processing industries etc. Indian politicians and critics were not aware of the problem.

Conclusion

The foreigners discussion aimed at the misunderstanding of the implications of WTO norms and exposing the reality. India is a reasonable ellicient producer of agricultural products. But the problem lies with the employment of the people who are displaced in agriculture. In this case instead of finding faults with the international law we have think about the best application of the law.

25

Role of E-Commerce in Agricultural Marketing in Andhra Pradesh

R. Siva Ramprasad
A. Sreeram

The paper aims at protraying the current status and future need of E-Commerce in agricultural sector in A.P. and suggest some remedial measures for the education, training and research in the area of marketing for the development of rural farming community. E-Commerce practices in agricultural marketing sector is attempted to be studies under three broad heads–Education, Training and Research. The study is largely based on the secondary data collected and compiled mainly by IIM, Ahmedabad, National Institute of Agriculture Marketing, Jaipur, Agriculture Marketing Training Centre, Nagpur; The Directorate of Marketing and Inspection.

Agricultural Marketing in A.P.

Agriculture Marketing scenerio in A.P. is completely in the clutches of middle-men and bureocracy forces. Government ad-hoc policies also exalrate the situation in the Marketing problems. All farming community in A.P. are in desperate mood about their respective profession, because of some rumors spread in entire A.P. in particularly paddy, cotton, pulses growers were fear about the import of China Agri-products to India with low prices. After March 31st another threat to the farmers is Australian milk products dumped into the Indian market. The current year the Govt. of A.P. is facing a lot of resistance from the farming community about the remunerative prices. Main cause was during 1998-99 rice and pulses per 100 kg bag Rs. 968, Rs. 2380 in 2000-2001; Rs. 805, Rs. 1400 respectively. In this context we require a well developed marketing information technology policy and frame-work that is also with the low cost paradigm, which is only possible through the E-Commerce Services.

What is E-Commerce?

According to WTO "Production, Distribution, Marketing, Sales or Delivery of goods and services by Electronic means".

Electronic Marketing

Electronic marketing is the sub-disciplne of E-Commerce. "EM is an interactivity, where a marketer establishes two-way communication with customer in what is called asynchronomous domain". EM encompass a range of technologies from promotional floppy disc to the interactive television store friends, to consumer on-line services.

E-Market Place

EMP is the digital environment or local where a marketer presents these promotional messages and where a potential buyer can purchase a marketer's products or services and requests more information.

E-Commerce Education

Eductaion facilities have been provided input marketing/ agriculture production aspects right from the level of the village level worker upto doctorate level. But output marketing has suffered a total negelect on educational fronts. According to survey report brought out by the Punjab Agricultural University, Ludhiana, the subject of the agricultural marketing forms only one of the topics included under the subject 'Economics' or 'Agricultural Economics'. The content of Agricultural Economics in the total package of agricultural courses in 1.57 per cent at B.Sc. (Agri.) general level and 2.34 per cent at B. Sc. (Argi.) Economics elective level for under-graduates. At the post-graduate level, the major content is that of economics plus 10-12 credit hours for supporting courses. Further, out of agricultural economics, the proportionate share of agricultural marketing was 17.19 per cent for the general scheme and 20.14 per cent for elective scheme for under-graduates and 17.15 per cent for post-graduates. At the doctorate level in agricultural economics, weightage for agricultural marketing was about 22 per cent only. E-Commerce in curriculum in the same level of education is absolutely nil except IIM, Ahmedabad. As result of such an utter neglect the personel employed in managing the marketing activities have been a mass of hetrogeneous mix. Hence, keeping in the mind agricultural sector need E-Commerce education.

As far as graduation and post-graduation level of education is concerned, it would aim at:

- Training graduates and post graduates for teaching, training and research in E-Commerce Education.
- Give training to upgrade the knowlegde of college and University teachers on E-Commerce and concern research activities.
- Providing digital environment infrastructure for interchange of ideas information regarding to E-market places, knowing price situation, agricultural marketing opportunities across the globe.

Post-graduation education E-commerce is one of the specialisation would especially be essential for the policy-makers, administrators, University teachers and professionals enagaged marketing research, planning and policy-making. The proposed E-Commerce package includes on environment analysis, quantative techniques use of computers in information management, marketing research, E-advertising, product promotion, EC infrastructure and security, the site server, platform server, side componments, web user interfaces, well developed IT companies E-Commerce solution case studies.

E-Agricultural Marketing Training

The DMI has been pioneer in offering the different training programes to agricultural marketing. On-line marketing point of view their contribution is negligible, keeping in the mind all universities and institutions must start the training programmes on self-finance bases. The objective of training at grassroot level would be:

- Improving market skills, operation and services, thereby reducing middlemen in the market chain.
- Adoption of computer education for upgrading the existing facilities and services for marketing agriculture products and food.

Such training programms sponsored by the Central/State marketing authorities and also through NABARD, Co-operatives and private training institutes. At middle level training programmes would aim at

- Upgrading the marketing techniques and skills of the target groups.
- Improving their capacity for analysing and interpreting various types of marketing information through the internet and WWW technologies.

- Improving the supervisor capacity and acquinent with new market trends and related policies through the online services.

E-Commerce Research on Agri-Business

As per the Agri-business research in our country is concerned IIM Ahmedabad only started research recently. ITC and WIPRO companies only concentrated in their respective products. IIFT, CMIE, DMI started research on the E-trading. Steps initiated recent years with the view to intergrating the Indian Economy with the rest of the world and the subsequent impacts of such measures make it rather imperative to make consumer preference surveys as a part and the marketing research programmes.

The future research programmes should ensure aim at developing improved marketing technologies to :

- Suit internal and external marketing conditions especially highly diversed bio-physical and socio-economic conditions.
- Provide solution for agricultural marketing through the E-Commerce.
- Generate employment for rural poor and landless through E-Commerce activities and services.
- Improve the skill and productivity of the farm personnel through on-line services.

Conclusion

Agricultural marketing in A.P. is characterised today by the involvement of large number functioners and duplications of functions in the process of distribution of produce from producers to the consumers. Hence, there is a greater need to adopt E-Commerce education, training and research for the development of farming community as well as agriculture marketing in Andhra Pradesh.